THE MAGICAL WORLD OF

MOSS GARDENING

THE MAGICAL WORLD OF MOSS GARDENING

Annie Martin

TIMBER PRESS
Portland, Oregon

Page 1: *Fissidens dubius*
Pages 2–3: Welcome to my moss garden in Pisgah Forest, North Carolina, which offers year-round green pleasure.
Pages 4–5: Mosses revealed in melting snow
Pages 6–7, clockwise from top: *Climacium* species with spore structures, *Funaria* sporophytes arising from charred wood, *Atrichum undulatum*, male *Polytrichum* cups

Published in 2015 by Timber Press, Inc.

The Haseltine Building
133 S.W. Second Avenue, Suite 450
Portland, Oregon 97204-3527
timberpress.com

Printed in China
Third printing 2016

Text design by Laura Shaw Design, Inc.
Cover design by Laken Wright

Library of Congress Cataloging-in-Publication Data

Martin, Annie, 1953- author.
The magical world of moss gardening / Annie Martin. — First edition.
pages cm
Includes bibliographical references and index.
ISBN 978-1-60469-560-1 — ISBN 978-1-60469-647-9 1. Moss gardening. 2. Mosses. I. Title.
SB433.55.M37 2015
635.9'382 — dc23
2014048497

A catalog record for this book is also available from the British Library.

This book is dedicated to my dearest
brother and friend, Dale,
my beloved sons, Flint and Carson,
and my mossin' sidekick, Leila.

contents

INTRODUCTION

moss magic

Every time I walk out my front door, my mosses greet me, triggering a smile in my heart. The magic of my garden engulfs my spirit, and that internal smile instantaneously moves to my face. As I gaze at the dazzling mosaic of mosses in my garden, I feel immense joy and pride juxtaposed with calm. Throughout all seasons, my mosses have a magnetic quality that engulfs me. Most often, I am catching a quick glimpse for an infusion of moss magic as I head out into the world in my work as a moss landscape designer and moss farmer. What a divine way to start the day!

As I ride in my mossin' truck along the French Broad River heading north, my moss radar is on and I spy my favorite plant growing in hummocks and carpets in the deep shade of rich cove forests, braving bright sunshine on steep hillsides, and providing glimmering red displays of sporophytes (reproductive structures equivalent to flowers) atop rocky precipices towering high above. As I pass through town, I marvel at emerald tendrils draping over the edges of rock walls and velvety green filling the cracks of sidewalks. I welcome the brief delay of a traffic light to notice some heat-tolerant species thriving on a roof and remember the time I climbed up on that roof to retrieve samples of *Ceratodon* and *Entodon* species. While often associated with the natural environment of forests and waterfalls, mosses are common in urban settings, too. To me, there is something magical in plants that can live in such dramatically different environmental conditions.

The magic of moss gardening may start with the way it makes me feel, but many other aspects of moss fuel my love affair. Mosses offer visual delights and tactile pleasures; they stoke our imagination and our memories; they connect us to earlier peoples and moss gardeners around the globe; they have countless environmental advantages and medicinal uses; and they are simply a joy to garden with. For all these reasons, moss gardening is gaining in popularity. As more people become aware of the advantages of this alternative horticultural option, I am convinced that mosses will assume a place of respect within the gardening community. Mosses are on the verge of a verdant victory.

PAGE 8 Tolerant of cold and high heat, *Bryum argenteum* is commonly called sidewalk moss. Note the silver tips.

OPPOSITE *Dicranum scoparium* mosses provide incredible interest in the garden with their textures, shapes, and colors.

TOP My garden, with a myriad of moss species with native ferns, captures me with its magic every time I lay eyes on it.

In my garden, contrasts in scale and texture are provided by a granite boulder featuring *Polystichum* (Christmas) ferns to complement *Thuidium*, *Hypnum*, and *Polytrichum* mosses.

Not always green, some mosses such as these *Hypnum* and *Thuidium* species at the Southern Highlands Reserve in North Carolina turn a bright golden color during seasons with scant leaf canopy.

TOP This moss rainbow emphasizes various moss textures and colors.

Spectacular *Ceratodon purpureus* sporophytes (spore-producing plant structures) provide crimson displays above the velvety green of the colonies.

Male cups (part of the reproductive process) of *Atrichum* species offer an array of colors—orange, bronze, green.

VISUAL DELIGHTS AND TACTILE PLEASURES

Mosses have tremendous visual appeal, offering luxuriant expanses that encompass a range of colors and a variety of textures. Horizontal growers blend together to form lush carpets tinged with variegated hues. Mounding species of upright growers range in size from small pincushions to as large as a basketball. Green may dominate a moss landscape, but you can expect brilliant jewel-tone colors—crimson red, pale pink, tangerine orange, and sunshine yellow—to burst forth during the reproductive stage, providing stunning contrasts.

The term *moss green* first appeared in print in 1884, but with the range of colors displayed by moss species, I can't imagine which variation was given this designation. From a distance, expanses of mosses may blend into a homogenous whole. When you move in for a closer view, though, you begin to distinguish an array of tints. Shades of green reflect a tonal range and intensity of color that are a distinctive hallmark of these diminutive plants. As your visual acuity becomes attuned to the nuances, you notice that warm (yellow) and cool (blue) undertones distinguish different species. You discover mosses that are brilliant like gemstones—emerald, jade, and peridot. Others come in colors that can be associated with fruits and vegetables (Granny Smith apples, limes, avocados, and olives) or with crayons (yellow-green, medium green, forest green, and blue-green). Some species do not even seem to be green at all until examined closely. *Grimmia* and *Andreaea* species are such a deep, dark green that most of the time these mosses appear to be black. Even within the same family of mosses, *Bryum argenteum* seems consistently swathed in a veil of silver while *Bryum minutum* goes through a blazing red stage. *Sphagnum magellanicum* is a rosy, pale red color, and *Sphagnum palustre* can range from green to bronze to white.

Some changes in color indicate stress, but often mosses such as this *Aulacomnium palustre* will rebound within a few weeks or months.

Not all mosses are green all the time. In fact, you should expect differences to occur as a result of changes in moisture, sun exposure, or reproductive stages. *Hypnum* and *Thuidium* species regularly shift out of their green garb into bright yellow suits when the leaf canopy is barren or the sun exposure more intense. Many species tend to be greener when growing in deep shade. The opposite can ensue in sun, with golden overtones overpowering the base green as in the case of *Entodon* species. Additionally, during the life cycle of mosses, colors may be burnt umber, sienna, and yellow ochre. *Ceratodon* species can be bright green during the hottest of conditions, but after spore-producing stages, the plants turn brick red-brown. Soon regeneration happens and new tips of green spring forth from this dull, dingy state. Other mosses of a single species can display a rainbow of color variations within their life cycle, from glorious green to crimson and bronze colors with hints of orange and chartreuse.

Moisture heightens the translucent properties of glistening leaves and sparkling sporophytes, which reflect a prism of color through the droplets of dew that have accumulated during the night. Some moss species seem luminous because the thin leaves transmit light. An individual leaf of moss in *Plagiomnium* species may appear light green, yet the whole colony may emit a neon green glow. While moist, the upper leaves of *Rhodobryum* create the appearance of green rosettes or roses. When sun hits the moss, its leaves shrivel and curl up, becoming almost unnoticeable in the landscape. Upon rehydration, the moss magically transforms from a brown nubbin into its blossomlike appearance within minutes. In my own garden, I feel angst when mosses dry out, and I obsessively respond to my compelling desire to give them a rejuvenating drink. As they begin the saturation process, I regain my own glowing state as I watch leaves swiftly unfold and colors magically intensify.

Mosses can be mesmerizing. A closer look into the micro-mini world reserved for botanists, bryologists, and moss lovers like you and me reveals fascinating details. With the use of a handheld lens or jeweler's loupe, you can see distinctive differences in colors, textures, shapes, and sizes. The magnified view of neon leaves of *Plagiomnium* species gives off a dazzling luminosity; I can see how the leaves form shapes to collect extra water droplets and moisture. The petite leaves of *Climacium* species have a radiance that truly rivals fine emeralds. Checking out the details of leaf shapes or spore capsules with a hand lens yields exceptional enchantment in exponential doses.

The tactile sensory aspects of mosses rekindle our direct connection with nature. It soothes the spirit to touch mosses and experience the coolness upon our fingertips. We want to touch the tufts of green with our hands and walk barefoot on the lush carpets. Our feet delight in the variations in how spongy, thick, firm, or lofty different species feel. For me, the stimulus of the senses raises my spirits and improves my attitude. And walking on mosses is actually good for them, helping them attach to the surfaces where they are growing.

The magical ambience of mosses in moonlight is yet another visual and tactile pleasure. Many times I have danced around like Isadora Duncan, singing the lyrics to a Van Morrison song, "It's a marvelous night for a moondance." Being a free spirit dancing barefoot in my moss garden is a fantastic way to experience full moon fever!

IMAGINATION AND MEMORY TRIGGERS

A moss garden serves as an imagination station for children as well as adults, and many memories of mosses are rooted in childhood experiences. As people share their own stories

Peering into the micro-mini world of moss colors, textures, shapes, and sizes with a handheld lens yields exceptional enchantment.

Moss mats planted contiguously provide instant gratification and beg to be walked on.

at my lectures and workshops, they beam with joy. Integrated with positive associations with places and loved ones, their moss moments are transfixed in their minds as if suspended in time. Green imagery is predominant and pleasure is a central component. Imaginary friends and fairies are described as playmates. Moss carpets and embellishments fashioned from lichen sticks, pinecones, and clover garlands (arches, swings, and tiny chairs) are common themes.

Carole recounted her treasured memory of a moss garden known in childhood and seemed to regain that elusive feeling of uninhibited imagination and childhood wonderment. Thirty-something Sam came up after a talk and with some embarrassment told me that as a child when he visited his grandmother's home in the summers, he would sneak off to the side garden with an inviting green moss carpet, take off his clothes, and roll around naked. The coolness was just too hard to resist on a hot summer day for this little boy. Ah, the innocence of childhood. Now that's throwing all inhibitions to the wind, isn't it?

Paula Steichen, granddaughter of respected author Carl Sandburg, recounts her own moss experience in her book *My Connemara*. She recalls, "I planted a moss garden by his work area on the rock [a granite precipice where Sandburg liked to sit and write], and I often appeared by his side, asking him to come and see a sight in the garden—brilliant crimson spores or a yellow toadstool." Her memories of mosses are intertwined with a love of nature that she shared with Buppong (as she called her famous granddaddy). Do you have your own moss memory? Can you still feel the magic?

My own first memory is vague. Mosses seem to have always been in my life whether I was conscious of them or not. Growing up in the mountains of North Carolina, I played outdoors quite a bit and our family went camping in the forests, so bryophytes (plants with little or no vascular tissue, which includes mosses, liverworts, and hornworts) have been ever present. My definitive remembrance is when I made my first terrarium at about ten years old for my pet chameleon. While I approached this project with creativity, I had a functional purpose in mind—to provide an inviting environment for Oscar. Since I had noticed other lizards, frogs, and salamanders living in mosses in natural settings, I figured he would like them in his terrarium home. I did my first moss gathering to create this miniature world, little suspecting that it was the first step on my journey to becoming a moss artist.

MOSS APPEAL ACROSS TIME AND SPACE

For me, gazing into the infinity of moss plants is akin to reaching across the millennia of time when stargazing. I realize that my time on earth is a brief encounter in comparison to these ageless gems. The magnitude of millions of years of survival elevates them to a state of gentle giants, and it is I who feel like a transient occupant on this planet. Insights into the meaning of our existence are among the intangible rewards that the glories of mosses bring to our hearts and minds.

Moss appeal transcends time. There is no doubt in my mind that the intricacies of translucent leaves and the glimmering sparkle of sporophytes caught the attention of others who have come before me on this planet. I wonder whether primitive men and women appreciated the delicate beauty or if they used mosses only for functional reasons in their lives, perhaps sleeping on comfortable, cushiony moss beds in the safety of their cave dwellings. With the looming issue of daily survival, creating gardens was probably not a high priority for our earliest ancestors. When lifestyles became more sedentary, several types of mosses (*Floribundaria*, *Sphagnum*, *Thuidium*) were used for chinking between logs to stave off cold winds and deaden sound, and moss was burned for fuel to keep warm. *Polytrichum* plants have been intertwined into nautical ropes. It is not known when these minuscule green plants first appeared in gardens. Probably they sneaked in—just as in today's gardens, where spores and fragments are transported by the wind, rain, birds, and critters.

A love of mosses crosses cultural boundaries to unite "nature's chosen ones," a term coined by George Schenk, author of *Moss Gardening*. I feel connected to other gardeners around the world who treasure these delightful plants in the way I do. Surely they must experience similar feelings of elation. In fact, I feel an alliance with enthusiasts on the other side of the globe. I have the greatest esteem for the dedicated stewards who have tended mosses in the impressive temple gardens of Japan, bestowing these precious gifts on subsequent generations.

THE JOYS OF MOSS GARDENING

Gardeners of all sorts know the pleasures derived from the actual process of gardening—planning, planting, and nurturing. And so it is with mosses as well. Maybe I am a bit prejudiced, but it seems to me that these minuscule plants magnify the joys of gardening. In many ways, you do not have to work as hard to reap the benefits.

For one thing, you won't get an aching back from digging holes, a common task associated with other kinds of gardening. Soil prep is minimal. If you have acidic soil and choose acid-loving mosses, you are good to go. If you have more alkaline soil (predominant in urban locations), you can choose moss species that prefer alkaline conditions. Of course, you can modify soil to achieve your preferred soil pH (from a French term, *pouvoir hydrogen*, "hydrogen power"). Even if you have soil so inferior that even weeds avoid it—sandy soil or red clay or even gravel—you do not need to amend it for many moss species. Since mosses do not have roots and absorb nutrition through their leaves, it does not matter to them if the soil is substandard. Do not waste your money or time by adding good garden soil. Moss colonies can be laid right on top of existing, atrocious soil substrates, and walked on to help rhizoids (rootlike filaments) attach.

Often gardeners are faced with locations where nothing seems to grow, especially in deep shade or in moist, boggy spots. Repeated efforts to plant grass can be frustrating. And yet

Mossin' terminology

You will find the terms *moss/mosses* and *bryophyte/bryophytes* used interchangeably as nouns throughout this book. Although the dictionary does not have a verb related to moss collection, moss gardening, and/or moss mania, *mossing* is a common colloquialism in many regions. Since we don't say our *g*'s in the South, I spell it with an apostrophe and omit the *g* (*mossin'*). The person who practices any moss activities is a *mosser*. The nursery where I cultivate mosses is a *mossery*. You are on your way to becoming a *mosser* if you are reading this book.

Here I am out mossin', retrieving mosses from asphalt roads at the invitation of the Sherwood Forest Property Owners' Association in Brevard, North Carolina.

those sneaky mosses seem to appear in these very spots out of nowhere. Eventually, some homeowners give in and accept that trying to grow anything else is a losing battle. If the area stays soggy, mosses can help resolve puddling issues. On steep hillsides with slopes that make maintenance difficult, mosses can hold the soil in place. The key is to choose the appropriate moss species for the site.

Unlike your neighbors who may take pride in their lawn and garden achievements only in the summer, you will enjoy your moss haven in the winter, too, since they do not die back and go through winter dormancy like other plants. If anything, they become even more attractive in contrast with the dormancy of many vascular plants. Trees and cold-hardy shrubs such as rhododendron really stand out when skirted with a carpet of moss plants.

Planting zones based on cold hardiness are irrelevant to mosses because they can tolerate subfreezing temperatures. Phenolic compounds make them resistant to the negative effects of cold weather. They do not croak when they freeze; encased in crystalline ice or blanketed in snow, mosses hang in there. In fact, upon thawing, my mosses look just as good as before their frozen state! As the weather warms and melting occurs, vivid green mosses are sprinkled with new sporophytes.

I must acknowledge that mosses are not low-maintenance plants. While you can eliminate some typical lawn and garden chores such as deadheading, weed eating, and mowing, you'll still need to water, weed, and remove debris. I adamantly advocate supplemental watering as a key factor in achieving long-term appeal. Some people say that mosses prevent weeds. I am happy for them. I can testify to dealing with weeds, tiny ones, way more than I would like. Gardeners who take these extra steps generally end up singing praises about the improved growth and beauty of their mosses.

ENVIRONMENTAL BENEFITS

Go green with mosses! We are bombarded with television commercials on how to go green in our gardens—but most of the time, environmentally unfriendly methods are recommended to achieve green, particularly the application of chemicals to promote growth, inhibit plant diseases, and eliminate insect pests. By contrast, moss gardening is truly green through and through. You do not need a chemical "green thumb" to succeed—no fertilizers, pesticides, or herbicides. By eliminating the use of poisons, you can play a valuable role in reducing groundwater contamination and runoff of hazardous chemicals into our natural water resources. When you stop mowing, you are taking personal steps to reduce air pollution. Mosses are an environmentally benign way to conserve water, control erosion, filter rainwater, clean up hazardous chemicals, and sequester carbon. Also, mosses serve a valuable ecological role as bioindicators for air pollution, acid rain, water pollution, and wastewater treatment.

All of these side benefits of beautiful bryophytes make them good eco-friendly choices for our gardens. Add the feathers of a responsible steward and champion for sustainable landscapes to your jaunty moss gardener's hat. Every little bit helps, and it all adds up to a better world.

No poisons or pollution

Many gardeners and landscapers rely on fertilizers, but as a moss gardener, you will need no soil supplements to encourage growth. In fact, too much nitrogen along with other macro, secondary, and trace nutrients could be harmful. Since mosses "eat" dust particles in such minute quantities, fertilizers could actually hamper moss growth. Even organic fertilizers might provide too much of a particular nutrient or micronutrient.

Beneficial insects, salamanders, frogs, and the like live in moss colonies. Because of internal anti-herbivory compounds, mosses taste bad to typical garden insect pests or critters. Deer might stomp around, but they don't eat mosses. Reindeer have been documented eating reindeer moss, but it is a lichen and not a true moss (bryophyte). Further, insects inhabiting mosses do not eat them or cause any significant damage. Therefore, pesticides are unnecessary.

To complete your shunning of the trio of harmful environmental substances, you can stop using any herbicides, too. Biochemical compounds in mosses function like antibiotics to deter any diseases, so mosses are not subject to the vast array of problematic diseases that plague other land plants. It is a rare occasion when mosses get any fungal disease or harmful pathogens. Although it is possible for mosses to get sick, in my experience it been the exception rather than the rule. When mold occurs, it is usually due to overwatering, lack of attention to debris, or multilayering of moss colonies as a result of critter shenanigans. As moss gardeners, we do need to be aware of problems and troubleshooting protocols. Information is becoming available in this area as scientists are beginning to research potential fungal issues regarding cultivated moss growth.

By planting mosses instead of a grass lawn, you can help reduce air pollution. Not all air pollution comes from industrial sites or big cities. Each week, millions of suburban grass lovers send carbon smoke signals into our atmosphere. If you are still cutting your grass with a gasoline-powered mower, you are contributing to this problem. Lawnmowers, weed eaters, and leaf blowers are only minimally regulated. They do not have catalytic converters, and that obnoxious smell is a harmful discharge. Mosses do not have to be mowed.

Water conservation

Because of low annual rainfall in arid regions, water supplies can be precious commodities. Watering restrictions are the norm. During

In places where nothing else will grow, mosses provide an outstanding year-round green solution.

RIGHT *Leucobryum glaucum* and *Dicranum scoparium* skirt this tree base.

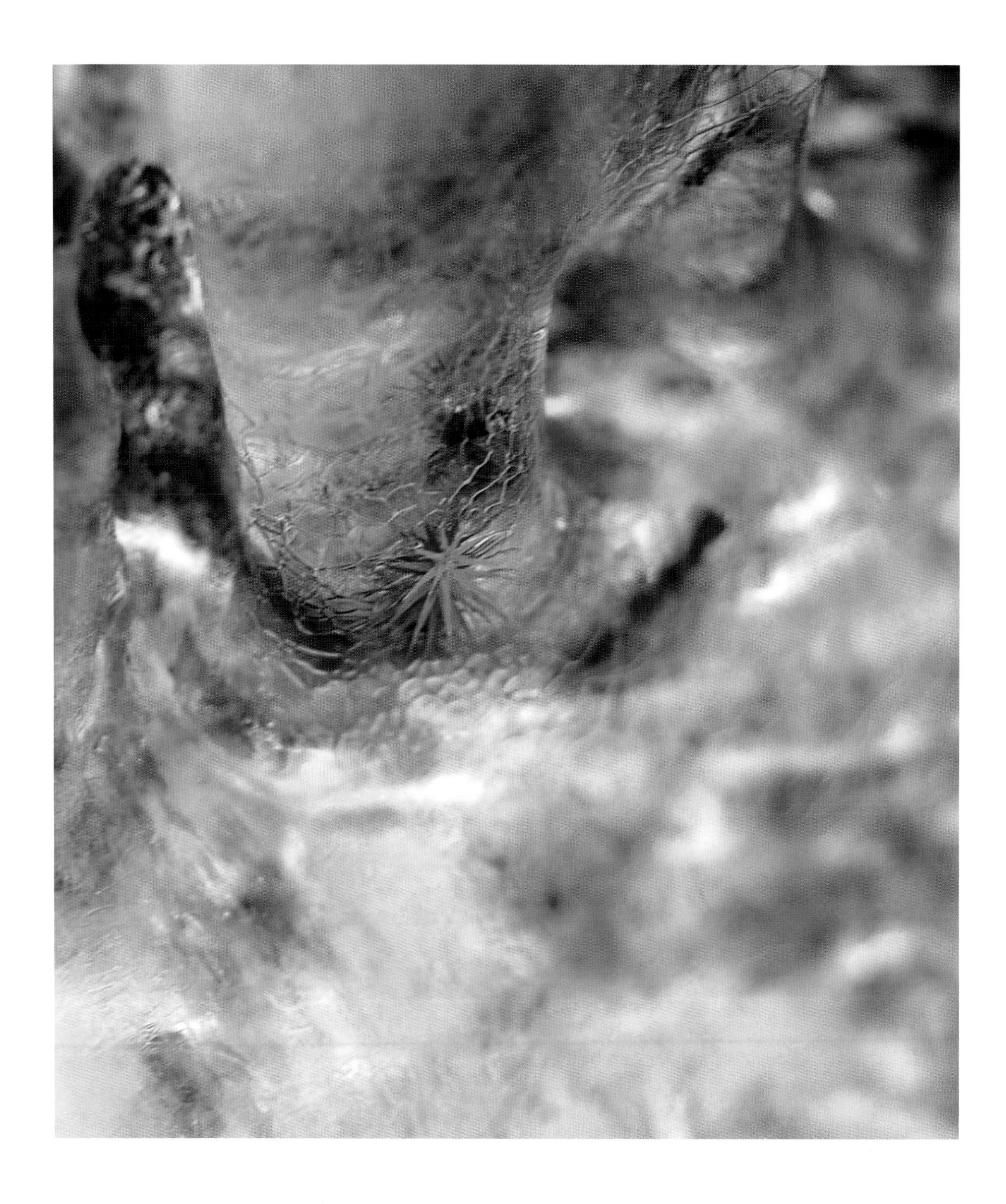

times of drought, some places may temporarily restrict watering for landscape purposes. Mosses do require moisture and thus are not xeriscaping solutions for landscapes; they do much better if watered regularly. But if you are concerned about water conservation, using rainwater collection systems and these tiny plants that require only brief, light watering sessions could be part of your solution. Rather than long drenching soaks weekly, mosses prefer short sessions each day. Misting irrigation systems that use a fraction of the water of other kinds of irrigation systems are enough to keep mosses happy.

Erosion control and flood mitigation

Many moss species are guardians of the soil and keep it from washing away. *Polytrichum commune,* an upright grower, can be planted on steep hillsides in nutrient-poor soil. This species is an excellent solution to erosion concerns even in sunny locations. Its long rhizoids steadfastly hold red clay, gravel, and sandy substrates in position. Steep, almost totally vertical slopes hold together as tiny *Pogonatum* mosses cling to exposed substrates. Rushing water slows down to enter the groundwater table gradually thanks to the absorptive properties of moss leaves.

The harsh appearance of riprap used to hold a bank in place can be softened by introducing *Thuidium delicatulum,* a fernlike sideways grower. Planting mosses in drainage ditches reduces the impact of stormwater runoff. And let's not forget our roofs—bryophytes can be featured in contemporary green roofs that insulate and cool, filter air pollutants, and reduce stormwater runoff.

In America, most newly constructed green roofs are composed of sedums or grasses, but mosses are used as part of green roofs in European countries, and in 2011 I created a roof at the North Carolina Arboretum as an example of how mosses are feasible on green roofs. Incorporating rainwater harvesting with a misting irrigation system, this green roof makes it possible for the mosses to do well even in direct sun. *Polytrichum* and *Ceratodon* species have been the best performers, with intense growth and several seasons of sporophytic displays, but I must acknowledge that *Leucobryum* species have struggled, so not all mosses are happy campers on rooftops. Mosses such as *Bryum argenteum*, *Ceratodon purpureus*, and *Hedwigia ciliata* do tolerate high heat and grow on many roofs in my region. In the Northwest, *Dicranoweisia cirrata*, *Bryum capillare*, *Rhytidiopsis robusta*, *Racomitrium canescens*, and *Tortula princeps* are abundant on roofs.

Filtration and phytoremediation

In terms of water filtration, *Sphagnum* species are especially effective as filtering and absorbing agents in the treatment of wastewater. Research indicates that mosses can be useful in addressing toxic discharge of undesirable elements (silver, copper, cadmium, mercury, iron, antimony, and lead). In urban areas, these advantages could play an important role in dealing with excessive rainfall, poor drainage, and flash flooding associated with nonpermeable surfaces. Further, mosses filter out organic substances such as oils, detergents, dyes, and microorganisms. In our streams and rivers, bryophytes are essential in food web interactions and nutrient cycling, contributing to the total stream metabolism.

Mosses can help reclaim land at abandoned sites where mining has occurred. Many species are tolerant of heavy metal toxins. *Scopelophila cataractae*, commonly referred to as copper moss, is often associated with copper deposits and can be the first species of plants to reappear without any special efforts, yet the beneficial effects of this category of pioneer plants may go unnoticed in formal efforts to restore forsaken landscapes.

In places where people brave extreme winter weather, salt is used extensively by road maintenance crews and homeowners digging out. As the snow melts, these salts are absorbed

OPPOSITE
This *Polytrichum commune* is temporarily encased in an icicle but will look just as good after thawing as before.

into the soil and penetrate into groundwater. Additionally, after a heavy storm, runoff water from nonpermeable surfaces rushes through urban drainage systems to lakes, rivers, wetlands, and coastal waters. According to the United States Geological Survey, "Stormwater picks up potential pollutants that may include sediment, nutrients (from lawn fertilizers), bacteria (from animal and human waste), pesticides (from lawn and garden chemicals), metals (from rooftops and roadways), and petroleum by-products (from leaking vehicles)." I am glad to report that mosses are part of the solution rather than being victims of pollution—they slow down stormwater while tolerating unfiltered contaminants. Sometimes I rescue mosses from places subject to runoff from roads and impervious asphalt parking lots. Despite exposure to oil residue from cars and trucks as well as seasonal road salt, bryophytes survive and even thrive.

Furthermore, mosses sneak in and grow at plant nurseries in spite of sanitation practices that include chlorination, bromine injections, and/or ozonation of irrigation water. Wouldn't you want plants that can handle even the worst of circumstances in your own yard?

Grass lawn obsession

An estimated 80 percent of American households have grass lawns. In terms of land use, turfgrass covers almost 28 million acres in the United States. Of this staggering number, 21 million acres are private lawns. North Americans spend more than $40 billion dollars each year to keep grass lawns looking good. It takes a lot of work—more than 150 hours per year (that's three entire work weeks)—for homeowners to maintain grass lawns. Yet, this obsession with turfgrass, with never-ending chores and lots of money from our pockets, comes at a significant cost to our environment.

Lawnmowers are responsible for 5 percent of the air pollution generated in the United States. Gasoline-powered lawn mowers emit ten to twelve times more hydrocarbons than automobiles and trucks. Weed eaters emit twenty-one times more. But leaf blowers take the pollution prize—thirty-four times more! Do the multiplication. It is an alarming amount. One hour of lawn mowing equals forty-three hours of driving around in a vehicle.

Lawns and golf courses deplete precious water resources, too. Keeping grass green takes 1 to 2 inches of supplemental watering each week. An estimate from the US Environmental Protection Agency suggests 30 to 60 percent of urban fresh water is used for grass lawns. Significant groundwater contamination comes from the use of fertilizers, pesticides, and herbicides used by American gardeners with expansive grass lawns.

You can be part of the solution to environmental issues if you choose moss landscapes over grass lawns. You can reduce air pollution by putting away gas-powered mowers, weed eaters, and blowers. You can reduce your groundwater contamination since mosses do not require any chemicals. You can conserve water resources with mosses (even if you do provide supplemental watering sessions). So, what do you think? Are you ready to say good-bye to grass lawns and hello to mosses?

Carbon sequestration

Mosses play a significant environmental role in the global carbon cycle as the largest land repository for carbon on the planet. *Sphagnum* peatlands soak up vast amounts of carbon dioxide from the air, far exceeding the rate of carbon sequestration by all rainforests combined. (When the rate of plant production in an ecosystem exceeds the rate of plant decomposition, carbon sequestration occurs.) Peatlands contain as much carbon as is present in Earth's entire atmosphere; it is estimated that they sequester between 198 and 502 billion tons of carbon.

You see, more than 60 percent of the world's wetlands are composed of *Sphagnum* mosses and decaying vascular plant matter, and *Sphagnum*-based bogs cover 2 to 3 percent of our planet's land mass—more total land mass than all the rest of the plants on Earth, including all other trees, grasses, and flowers, combined. *Sphagnum* may sequester more carbon than any other land plant, according to bryophyte ecologist Janice Glime, although aquatic algae do exceed *Sphagnum* in volume of carbon sequestered globally.

These globally significant ecosystems contain one third of the world's soil carbon and 10 percent of our freshwater resources.

Approximately 175 countries around the world have peatlands that serve as carbon sinks. In North America, large expanses of *Sphagnum* peat occur from the boreal forests of Canada to the Everglade's fens and swamps in Florida.

While Planet Earth's inhabitants could benefit from these moss enclaves, regrettably we are destroying them. The extensive devastation of peat bogs is happening due to a variety of factors including global warming, irresponsible harvesting in mass quantities, agricultural expansion, commercial exploitation, and fire. It is estimated that 7 percent of peatlands globally have been exploited by clearing and draining areas for agricultural endeavors. Indonesia has more tropical peatlands than any other nation but is losing them at an alarming rate—nearly 250,000 acres each year.

The loss of rainforests through massive clear-cutting is associated with degradation of air quality and loss of carbon sequestration capacity worldwide, and the loss of *Sphagnum* biomass may have a similarly negative impact on global air quality and climate change.

Billions of tons of methane, a greenhouse gas more potent than carbon dioxide, are being released as the world's largest peat bog (in Siberia), equal to the size of France and Germany combined, is thawing for the first time in eleven thousand years. And when peat bogs catch on fire, they can smolder for weeks, months, or years, releasing carbon dioxide all the while. The ominous cloud that covered Southeast Asia in 1997, attributed to Indonesia's burning wetlands, lasted for months.

The destruction and corresponding decline of peat bogs due to humans could be turned around with the application of best management practices to regrow mosses. Researchers at Bangor University in Wales are currently investigating how the growth rate might be increased to achieve sustainability.

COMMERCIAL USES

Speaking of commercial harvesting, mosses are globetrotters with all types of commercial uses. In Ireland, harvested peat is compressed

RIGHT *Thuidium delicatulum* covering a steep bank in the Vickery garden in Brevard, North Carolina, provides effective erosion control.

LEFT Mosses don't need to be mowed, thus avoiding the air pollution produced by mowers, and *Sphagnum* mosses actually clean our air.

After rescuing *Hedwigia*, *Ceratodon*, and *Entodon* species from another nearby roof, I put these mosses on my own roof.

BELOW I created this green roof featuring mosses atop a garden shed at the North Carolina Arboretum in Asheville, North Carolina.

into bricks to be used as a heat source; they burn slowly at a high temperature, providing welcome warmth on blustery, cold days and nights. In the British Isles, mosses are burned in the distillation of Scotch whiskey, giving distinctive smoked flavors to the barley. From east to west and north to south, mosses (*Sphagnum*, *Thuidium*, and *Neckera*, to name a few) are used for packing and shipping live plants, bulbs, and apples. The worldwide floriculture industry promotes the use of mosses (usually dried) in floral arrangements and potted plants. The popular method of propagation by air layering with mosses nurtures the growth of other plants.

The use of peat mosses in landscaping as a soil conditioner is commonplace, particularly in America. Small garden centers and big box stores sell thousands of cubic feet of peat each year to commercial landscapers and home gardeners. Little packets of dried mosses (*Hypnum*, *Dicranum*, and *Thuidium* species, and more) are sold for decorative floral purposes (to cover the base of potted plants and silk trees) and craft projects (wreaths, topiaries, baskets). Buying these dried mosses drives market demand, but it's hard to know if the mosses are harvested responsibly or commercially grown in a sustainable fashion. My recommendation is that you avoid buying any mosses sold dried or in bags.

MEDICINAL USES

Sphagnum mosses can absorb up to 33 percent of their weight in water. The absorbent properties of mosses have provided medicinal and hygienic options for people throughout history. Today in Germany, mosses notably are used for their sterile properties in DNA research. Quite a few moss genera—including *Bryum*, *Plagiomnium*, and *Sphagnum*—are credited with medicinal properties and have been used to treat the common cold, tumors, cancer, fungus, diarrhea, hypertension, leukemia, infections, cardiovascular diseases, fever, inflammation, and skin diseases.

Mosses, and their companion species liverworts, have been ingredients of herbal remedies for centuries. Native American women chewed the stems of *Polytrichum* mosses to ease the pain of childbirth. Historical records indicate indigenous people of various geographic regions of North America concocted brews and ointments from leaves as treatment for fevers, aches, and pains. Many other cultures, particularly the Chinese, have treated ailments with teas and poultices composed of bryophyte species. Compresses of *Plagiomnium* species can soothe blood blisters. *Bryum* and *Fissidens* species have been ground up and smeared on bald heads with the hope of hair regrowth. Various Native American tribes used *Sphagnum* mosses as baby diapers and for the periodic needs of women. In 1991, Johnson & Johnson introduced moss sanitary napkins in Canada, but the product never caught on.

Naturalists, wilderness adventurers, and mossers can apply a tincture of *Papillaria* and *Thuidium* species as an insect repellent. You can even stuff mosses into your hiking boots to curb sweaty feet. If you are wounded on the trail, find some *Sphagnum* moss to dress the wound. During World War I, nurses and doctors discovered wounds healed faster with moss dressings than the standard, yet scarce, cotton bandages. *Sphagnum* bandages are still in use in China today.

SPIRITUAL INFUSION

In addition to the many practical reasons to love mosses and to want your own moss garden, there is the fact that immersing yourself in the tranquility of a moss garden will enrich your spirit and improve your quality of life. Moss landscapes offer verdant vistas that seem to infuse our spirits with a sense of serenity. The breathtaking beauty of a forest floor carpeted in green with beguiling moss-covered rocks strewn along a mountain stream always rejuvenates my spirit, grounding and centering me with the rhythms of the universe. My moss garden offers an escape from the hustle-bustle of daily living—a place to shed a tear or two, sing my heart out, or commune with nature. In my haven, I not only contemplate the meaning of my existence amid these ancient plants, but I also revel in the natural forces that connect humanity with our place on Earth. My spirit soars in my moss garden. Just being among the mosses helps me regroup and regain my positive attitude toward life's challenges and struggles.

Since childhood, I've had a connection with mosses. I am not sure whether the mosses chose

Sphagnum peat bogs make up 2 to 3 percent of the world's land mass and serve as the largest carbon sequestration sinks on land.

OPPOSITE What was once a driveway is now a serene green retreat at my home in Pisgah Forest, North Carolina.

me or I chose the mosses, but I now realize that I was destined to be a moss artist. Deep within my inner being, I intuitively know that my path to enlightenment is a mossy one. With an innate sense of moss design and passionate desire, I am committed to creating impressive public and private moss gardens. I truly believe that my purpose in life is to share the joy of mosses with others. This book is my vehicle to reach out and impart my expertise to you. As you continue reading, you will find that the goal of this book is to provide inspiration, environmental justification, and practical advice on how to achieve moss magic in landscapes. I hope you will discover your own intangible rewards as you enter the magical world of moss gardening.

green grandeur

A TOUR OF MOSS GARDENS

Moss gardening has been a mainstay of Japanese gardens for centuries. More recently, the charismatic charm of marvelous mosses has caught the attention of gardeners on this side of the globe, too. Despite the preoccupation with grass lawns and formal color gardens, some Westerners are adopting the attitude that mosses belong in their landscapes. Even Prince Charles has decided to welcome mosses to the grounds of Highgrove House. In Great Britain, mosses have been considered annoyances in grass lawns for decades. But His Royal Highness has stepped away from this conventional attitude by introducing a moss lawn in The Stumpery.

PAGE 30 The tops of *Rhodobryum ontariense* resemble green florets, but mosses have no true flowers.

OPPOSITE In Japan, serene temples called *koke-dera* (moss temples) draw thousands of visitors each year. Saiho-ji in Kyoto exemplifies these historical retreats with its appealing expanses of *Leucobryum* moss species.

While moss gardens are few and far between in the Western Hemisphere, they have been around for decades. Scientific works mention moss gardens existing in America as early as the 1930s. With grounds designed by Frederick Law Olmsted, the estate of William Bayard Cutting (now the Bayard Cutting Arboretum) on Long Island, New York, featured a moss garden. Historically, our moss garden mentors have either allowed bryophytes to creep in, taken measures to encourage this natural encroachment, or sometimes purposely introduced them. These early leaders entered the moss world with vigor and determination to create their own special refuges. Magnificent moss gardens are the legacy of these forerunners.

This chapter highlights a selection of premier public and private gardens. These featured gardens are representative examples of mosses growing successfully in a variety of landscapes and geographic regions. It is by no means an exhaustive list. There are many places worth visiting that are not mentioned.

If you do visit moss gardens in your travels, you will probably want photographs to commemorate your visit. Besides waving back at the camera, I hope you will take close-up images of moss species. Identification of types will benefit your own learning process. Pay attention to

The placement of stones and boulders is significant in Japanese gardens, as illustrated at Komyo-in, a subtemple of Tofuku-ji Temple in Kyoto, Japan.

placement of mosses in the overall design. Be particularly aware of environmental factors that may have a positive or negative impact. Note any techniques used by professional horticulturists and community volunteers that aid in maintaining beautiful moss features. Above all, don't rush through the gardens. Take time to absorb the phenomenon of living in synchronization with nature and to listen to the ancient moss whispers.

TEMPLE GARDENS OF JAPAN

While Westerners have been slow to embrace the beauty of bryophytes, the Zen Buddhist monks and their predecessors have nurtured moss havens for hundreds of years. The grand temple gardens of Japan represent the epitome of the world's magnificent moss gardens today. Historically reserved for royal families, elite society, hereditary military governors (shoguns), and priests, these splendid gardens are now open to the public. Starting in the twentieth century, thousands of visitors from all around the world have made the pilgrimage to visit Japan's national treasures. Entrance to these holy places requires an official permit, and numbers are limited. Before entering the inner sanctum, garden pilgrims slow down from the hurried pace of life through a period of quiet meditation and careful transcription of symbols with brush and ink.

As you begin your stroll in these serene and sacred gardens, a sense of timelessness consumes you. Simple in design, each *kokedera* (moss temple) reflects an unsurpassed connection with nature. Elements found in early Korean and Chinese gardens are evident, along with distinctive and stylized components that emerged during different eras of Japanese culture. Icons and elements reflect various religious philosophies—Shinto, Buddhism, Taoism, Hinduism, and Christianity. Shinto's "dignitaries of nature"—rock formations, ancient trees, and lakes—have left an indelible footprint. Components common to other religious beliefs such as islands, mountains, and seas blend with these earlier elements of design.

The gardens are dynamic and ever changing. Over the years, substantial transformations have occurred as a result of human influence and the impact of natural forces, including flooding. Out of necessity, perhaps as a result of neglect, many Japanese gardens have undergone major restoration in modern times.

In fact, only in recent times have mosses evolved as a prominent feature of the temples. It is believed that mosses were not included in original designs, but rather native moss species introduced themselves into the gardens over the years, particularly during the Meiji period (1868–1912 AD). Benign neglect set the stage for moss pioneers to move in and take up residence.

Maintenance of moss gardens includes meticulous sweeping.

Weather impacts the beauty of moss gardens. When rainfall is plentiful, they look magnificent, but during times of drought, even Japanese temple gardens like this one at Tofuku-ji Temple in Kyoto show signs of stress or dormancy.

Instead of resisting moss invasions, the temple garden stewards adopted these natural visitors, encouraging their growth and maintaining a favorable environment. Now mosses dominate the landscape. Japanese gardeners tend to the mosses daily, sweeping away debris with bamboo brooms and carefully removing weeds by hand. With reverence for these minuscule plants, the gardeners maintain extensive expanses of moss carpet.

The high humidity of the island nation provides an environment conducive to thriving mosses. However, mosses get hot and dry out at times even in the best of climates. Although glowing accounts of Japanese gardens highlight their glorious green, there are bound to be occasions when these gardens are not at their peak of beauty. If your visit is not during the rainy season or on a day when a shower has doused the landscape, the intensity of color may be less than you expected. One sightseer admitted, "I never observed any watering systems in the many gardens I saw in Japan. And they were often dry, so dormant." Do not be fooled that somehow Japanese gardens are not affected by climatic conditions.

Since these impressive gardens are located in Zen Buddhist temples, we often associate them with Zen thought. These tranquil retreats certainly provide inviting visual spaces for contemplation. Yet meditation is an internal experience with no requirement of external stimuli. In Zen philosophy, it is the creative process of gardening and the ongoing labor of maintaining garden grandeur that facilitates one's journey on the path to enlightenment. For moss gardeners, planting an interesting design and the ongoing labor of weeding, watering, and litter removal can serve as our meditative path.

Saiho-ji

One of the temples most famed for its moss garden is Saiho-ji in Kyoto. Originally built as an imperial garden by Prince Shotoku Taishi during the Asuka period (574–622 AD), it went through several transformations as a retreat for the royal family and distinguished aristocrats. The Ogon-chi (Golden Pond) with its islands and peninsulas was added during a time when floating on water was a preferred method of garden viewing and recreational boating was a popular pastime. But by the beginning of the Muromachi period (1336–1573 AD), a series of floods and fires as well as wars among the shogunate had changed the face of this garden, and it had fallen into a neglected state. To revive the garden, Muso Soske, a Buddhist monk and poet renowned for his horticultural skills, became the official head priest and garden designer at Saiho-ji. He renamed the garden to reflect Zen Buddhist concepts and used characters that mean "west" and "fragrance."

Muso's contributions reflected his desire to elevate a pleasure garden to a place of spiritual renewal—a temple garden. He placed emphasis on contrasts of light and shade and effectively incorporated opposites of mass and void. New perspectives of the garden are visible at every turn. Symbolic icons abound, with boulders mimicking mountains and meticulously raked sand and gravel areas serving as representations of water bodies. Longevity of life and prosperity is denoted by Kame-jima, a turtle made of stones, swimming in mosses symbolizing water. Structures with movable walls and square and round windows provide portals for viewing the majestic moss garden from all angles from indoors or outside.

Originally, only priests, Zen scholars, and honored guests were invited into the upper rock garden at Saiho-ji. The rugged approach of forty-nine steps, referred to as the Kojo-kan Barrier, signifies the journey to enlightenment and is not an easy passage. During the climb to this meditation garden, one loses sight of the lower garden as part of the spiritual experience. Perhaps Muso included this feature of large stone steps to signify his own journey to the summit of Mount Koin-Zan. A strategically placed sitting stone, Zaren-shi, provides a dedicated spot for solitary meditation with nature. Another early example of Muso's use of *kare san sui* (rock formations) is a dry cascade waterfall.

As many as 120 species of bryophytes grace the grounds of the Saiho-ji temple. Visitors immediately become immersed in the greenness of mosses and liverworts in contrast with delicate cherry blossoms in the spring and brilliant maple leaves in the fall. *Goke* is the Japanese word that refers to all bryophytes (and all other small plants), not just moss species. *Sugi-goke* is the common name in Japan for *Polytrichum juniperinum* (moss) and *Seni-goke* is the Japanese name for *Marchantia polymorphus* (liverwort). Other moss genera found here include *Climacium*, *Dicranum*, *Hypnum*, *Racomitrium*, *Rhizogonium*, and *Sphagnum*. Some of the liverworts found include *Bazzania*, *Conocephalum*, and *Porella*. Of special interest: Ginkaku-ji Temple (the Silver Pavilion inspired by Muso's architecture at Saiho-ji) has graciously provided identification labels in an educational exhibit of moss species.

Other moss gardens in Kyoto

Numerous private spaces and other public locations in Kyoto and surrounding areas have impressive moss landscapes. Crowded into confined quarters in urban centers, many Japanese crave a bit of nature outside their doorsteps. Pocket gardens (*tsubo-niwa*) featuring miniature moss scenes are common sights in courtyards and at entrance gates of private homes. Hidden behind walls, some moss scapes are reserved for personal pleasure. Others are open to the public. At restaurants, diners may satisfy their appetites in engaging moss settings. Temple gardens showcasing mosses abound. For example, the Tofuki-ji Temple features square patches of moss species arranged in a geometric pattern. Art lovers will be doubly rewarded with moss artistry while visiting the Gyokudo Art Museum in Mitake, the Nezu Institute of Fine Art in Tokyo, and the Hakone Open Air Museum in Fuji-Hakone-Izu National Park.

Raked white sand and pebbles represent flowing water, while mosses and shrubs symbolize islands at Manshu-in Temple in Kyoto.

LEFT Moss features can be found throughout Japan, in pocket gardens at private homes and at restaurants offering serene scenes like this one.

A patina of mosses reflects the antiquity of stone lanterns.

Lessons from Japanese moss gardens

Japanese moss gardens provide inspiration, and as moss gardeners, we can learn much from their thoughtful placement of design elements in a simple fashion. Water symbolized by sand or gravel meticulously raked in patterns to emulate ripples and waves is a crucial design element reinforcing the affinity of the Japanese people with the sea that surrounds their island home. In early Japanese gardens, sacred islands were an ever-present element. The influence of Buddhism made mountain representations more common. Large stones or boulders grouped in threes symbolize Buddha flanked by two Bodhisattvas. The pine is the basic structural tree. Manicured bonsai pines along with plum, maple, and cherry trees provide character and height along with towering bamboo plants. A monochromatic color scheme—green—dominates the landscape. Simplicity (*sanso*) through omission is central to achieving the desired tranquility. Asymmetry, rather than symmetry, is preferred. Shapes are irregular and groupings of stones or trees are in odd, not even, numbers.

Beauty is a constant even in functional elements. The graceful curves of bridges and the artful arches of garden gates further add to the unified whole. Water basins and stone lanterns signify functional aspects of the teahouses. Washing hands is part of the Japanese tea ceremony. Although today purely ornamental, stone lanterns were first introduced in the seventh century to light footpaths. In later years, lanterns provided twinkling reflections along the edges of ponds for evening boating parties. With the advent of the tea ceremony, they served a practical purpose for illuminating stepping-stones at night leading to the teahouses.

You may choose to emulate Asian styles and establish your own Japanese tea garden. One of the greatest lessons to be learned from these venerable gardens is the connection between nature and the human spirit while you *create* your moss garden. Patience, diligence, attention to detail, adherence to quality control, and a harmonious state of being are among the character attributes you may gain from your moss gardening experience.

PUBLIC GARDENS OF THE WEST

One might expect to find mosses in the botanical gardens of America. Following the inspirational stimulus of Asian influences, Japanese-style tea gardens are scattered from coast to coast—in San Francisco, Portland, San Antonio, St. Louis, New York, Durham, and Birmingham, to name a few locations. Often designed by horticulturists with close ties to Japan, these gracious gardens boast traditional architectural garden elements along with plants and trees imported from the Far East. When entering though the arched gates, visitors seem to step back in time to experience the serenity of Asian culture.

But surprisingly, as a general rule, mosses have not been prioritized as an essential element in Japanese-style garden design. In lists of selected plants, mosses do not rank very high and might not even appear in the official plant inventory at all. Since the horticultural value of bryophytes rarely is addressed in the curricula of our nation's best universities, graduates trained in the identification and care of vascular plants may not be knowledgeable about gardening with mosses. With the challenges of legitimately acquiring moss species and limited understanding of how to nurture their growth, botanical gardens have used alternative ground covers such as Irish moss (not a real moss) and grass lawns.

Yet even without intentional planting, mosses tenaciously invite themselves to take up residence and add their own special emerald elegance. Unfortunately, because many public gardens with restricted budgets and limited staff struggle to keep up with maintenance chores, mosses often take a back seat to other prized species and significant features. Weeds can quickly overrun mossy areas, and mosses may struggle without proper care and suffer from the lack of appropriate watering regimes.

As more people become intrigued with the exceptional beauty and environmental benefits of mosses, I am optimistic that we will see an upsurge in the creation of true moss gardens—in both the private and public sectors. The influence of Japanese style will continue, but innovative landscape designers and industrious homeowners can add their own personal flair, using contemporary approaches to featuring mosses in woodland gardens and native restorations, waterfall and pond surrounds, and outdoor living spaces.

Japanese Tea Garden, Golden Gate Park, San Francisco, California

Golden Gate Park in San Francisco has the notable distinction of housing America's first Japanese tea garden. Built in 1894 as an exhibit for the World's Fair, this garden was nurtured for several decades by Makoto Hagiwara and his family. It flourished under their meticulous care until these gentle Japanese-Americans were forced to leave their serene setting to live in a World War II internment camp. Falling into neglect during this dark time in American history, the gardens were later restored so that the legacy of the Hagiwara family continues. Today, thousands of visitors throng to this historic haven where mosses can be found growing on the walls of reflecting pools.

The climate of the Northwest supports moss growth even during the warm months of July and August.

OPPOSITE Mosses dot the landscape throughout the Portland Japanese Garden in Portland, Oregon.

Portland Japanese Garden, Portland, Oregon

Recognized as one of the best Japanese-style gardens in America, the Portland Japanese Garden encompasses 5.5 acres in Washington Park high above the city. Influences from Shinto, Buddhist, and Taoist philosophies infuse this tranquil retreat, where the connection of nature to the human spirit is a universal theme. The blend of common elements—stone, water, and plants—unifies five distinctively different garden areas. Design of the gardens by Takuma Tono, one of the foremost landscape architects of his time, began in 1963. After several reconfigurations and renewals, these exemplary gardens are now a popular tourist destination.

In one of the quintet of gardens, the Natural Garden (originally the Hillside Garden), Tono incorporated mosses into his original design. This garden has been renovated in recent years, and its design is considered the most contemporary of the Asian styles. The use of indigenous maple trees represents a departure from traditional Japanese plants; Tono felt that the garden should emulate the Oregon forests surrounding it. The mosses intentionally planted decades ago have suffered over time, but remnants of these mosses along with an abundance of volunteer species benefit from the moist Portland climate.

Today, a new grand master of Japanese gardening, Sadafumi Uchiyama, supervises garden care. As a third generation gardener, he possesses skills passed from his grandfather while he was growing up in Japan. As horticultural curator, he has trained garden crews and volunteers in proper ways to maintain garden beauty. As you wind through this walking garden, take time to reflect and contemplate life in this restful retreat.

The Bloedel Reserve, Bainbridge Island, Washington

"Harmony, respect, tranquility—how many times these very words have come to mind as I have walked about The Reserve." Founder Prentice Bloedel's declarations echo his intention to connect people with the natural beauty of Pacific Northwest landscapes at the Bloedel Reserve on Bainbridge Island, Washington. Attuned to the value of nature to the human spirit, Bloedel recognized that emotions of garden visitors could range from sublime calm to extreme exhilaration. In this 150-acre woodland retreat with landscaped gardens, you will discover the Japanese Garden, the Reflection Pool, and the Moss Garden, which is considered the largest in North America.

Bloedel was ahead of his time. As a twentieth-century pioneer in sustainability, he used sawdust to power his lumber mills and started a program of reforestation in clear-cut areas destroyed by logging. With a vision in his mind, he began to develop a retreat of monumental proportions with the purchase of this property in 1950. Originally a private estate, it opened to the public in 1988. Although renowned Northwest landscape design architects along with Japanese experts Fugitaro Kubota and Richard Yamasaki were occasionally involved, the gardens are really Bloedel's expression of his own sense of design and personal philosophy of life. He wanted visitors to experience nature in a way that could lead to a sense of enlightenment in this sanctuary. For more than three decades, these gardens were his most important life's work and culmination of his own dream.

As for mosses, natural colonization has occurred and the successional parade of species continues. The misty climate of Puget Sound near Seattle provides an ideal environment for growing moss. Frequent rain and harsh winds may play havoc with other plants, but garden workers report that the mosses thrive. In fact, all the wet weather in winter creates conditions that enhance moss reproduction. A newspaper article about the impact of severe weather described Bloedel mosses as "happy campers."

With more than forty species of bryophytes, the Moss Garden is a medley of texture, a bejeweled carpet of greens and yellows sparkling in the morning dew. *Atrichum selwynii* stands out as deeper green medallions under the shade and comforting moisture of companion ferns, ringed by golden ribbons of the more sun-tolerant *Rhytidiadelphus squarrosus*. Subtle variations in soil moisture define the domains of these mosses; *Calliergonella cuspidata* thrives in the frequently waterlogged depression and is bordered by colonies of *Atrichum* species along with *Polytrichum* species and *Leucolepis acanthoneuron* growing atop the tiniest of mountains. Moist rotting logs are claimed by *Plagiothecium undulatum* and *Tetraphis pellucida*. *Kindbergia oregana*, *Rhytidiadelphus loreus*, and *R. triquetrus* adorn weathered tree stumps, while *Racomitrium* species claim positions on boulders and are arrayed there as regally as a king on his throne.

Some species are native to the Pacific Northwest while others commonly live in my neck of the woods in North Carolina. There is a high degree of overlap, according to my bryologist/botanist friend James Wood. Logs and stumps are swathed in the kelly-green glory of *Dicranum scoparium* and *Tetraphis pellucida*—familiar species to me. In sections that are soggy, along the edges of the pond, and in drainage areas, liverworts dominate. After the long rainy season in winter, ephemeral sporophytes rise above the dense greenness of *Conocephalum conicum*. During this brief reproductive interlude (two to three days each year), the whitish yet translucent stalks supporting the liverwort capsules in which spores are formed "resemble a glimmering fog covering a miniature landscape," in the words of Wood.

The wetness of winter supports reproductive processes, providing awesome sporophytic splendor each spring for a vast majority of moss and liverwort species. In summer, when rain is infrequent and temperatures are hot, the mosses bide their time. The use of an irrigation system helps maintain their picturesque presence. But it is in the winter season when these mosses are marvelous beyond belief. Moss lovers owe a debt of gratitude to the Bloedels for leaving this luscious legacy to us all.

Take the time to stroll and enjoy the more than forty species of bryophytes at the Bloedel Reserve.

BELOW The Bloedel Reserve on Bainbridge Island, Washington, provides an ideal environment for mosses to thrive.

The Butchart Gardens, Vancouver Island, British Columbia

The Butchart Gardens near Victoria on Vancouver Island, British Columbia, have a color palette and textures out of a Renoir painting. Millions of flowers, shrubs, and trees delight thousands of visitors each season. Six gardens include The Rose Garden, Jennie Butchart's favorite, and the Star Pond where Robert Butchart's beloved ducks from his ornamental bird collection once swam contentedly. This couple's determined efforts to transform their land began in the early 1900s, and the gardens continued to evolve into a world-renowned attraction under the leadership of their grandson Ian Ross and his family. Today Robin Lee Clarke, great-granddaughter of the Butcharts and current owner, shares this century-old garden with the public.

The conversion of the original limestone quarry into the magnificent Sunken Garden is a remarkable example of environmental vision and land reclamation. Mrs. Butchart had tons of soil hauled into the gaping quarry hole via horse and cart (no easy chore by hand). The stunning gardens evolved as her legacy. This family's environmental consciousness and efforts to recycle and reuse following responsible stewardship principles are a garden priority today. A tradition since the mid-1950s, summer concerts and firework extravaganzas as well as twinkling Christmas displays keep the gardens active year-round. Strolling around the world is possible as you navigate your way through the Mediterranean, Italian, and Japanese gardens.

It is in the Japanese Garden that moss enthusiasts will be richly rewarded. Mrs. Butchart created this Asian feature in 1906 with

An element found in Japanese-style gardens, this moss-covered stone lantern is a prominent feature at the Butchart Gardens on Vancouver Island, British Columbia.

BOTTOM Walkways under arbors guide visitors past golden mosses at the Butchart Gardens.

the assistance of a Japanese landscape expert, Isaburo Kishida. A blend of traditional design principles and her own special touch make this a peaceful place. Moss genera found all around the world live here, including *Polytrichum*; tufted and cushiony *Bartramia*, preferring the moisture of water features; velvety *Ceratodon* and densely dark, scruffy *Andreaea*, adorning boulders and rock walls; and treelike *Climacium*, growing in thick miniature forests on the woodland floor.

More than twenty species of mosses and liverworts grace the grounds. Most likely, since these bryophytes are native to the area, the mosses moved in of their own volition. No documents exist to indicate that mosses were ever planted intentionally. Yet these green delights are omnipresent at the Butchart Gardens. "Horticultural workers tend to periodic weeding needs and the inevitable repairs prompted by natural conditions or created by visitors," explains Graham Bell, public relations and marketing director. He continues, "Committed to a sustainable approach, we use irrigation from the ponds as well as rainwater traps from building roofs and parking lots to provide supplemental drinks to the tiniest of species in the Japanese Garden."

Duke Gardens, Durham, North Carolina

Mosses are not new to the 18-acre William Louis Culberson Asiatic Arboretum in Duke University's Sarah P. Duke Gardens. Located in Durham, North Carolina, the arboretum was established in 1984. Various mosses had already established a home beneath the shady canopy of a 180-year-old white oak tree of grandiose proportions. Although curator Paul Jones has always loved mosses, he credits an energetic volunteer, Barbara Kremen, with spurring an appreciation of these sneaksters and providing maintenance to encourage growth. Since the 1990s, the mosses have continued to spread along a steep path embankment, providing a pleasing encounter for strolling visitors. Most unfortunately, a freak burst of wind in a summer storm in 2010 toppled the mighty oak. Without the protection of its expansive shelter, some moss species began to stress. A new action plan for displaying mosses ensued.

Having guided this garden's growth since the beginning, Jones, along with horticulturist Michelle Rawlins, brainstormed and dreamed of a new and expanded moss garden featuring a good representation of the diversity of species native to the Piedmont region. They had an available resource for acquiring moss samples—the vast forests owned by Duke in Durham and Orange counties, as well as several neighbors and friends. *Thuidium, Climacium, Atrichum, Polytrichum*, and *Rhodobryum* species, among others, were available for retrieval and relocation. *Bryoandersonia* ranks as the most prevalent of the thirty-something types of bryophytes—mosses, liverworts, and

Established in 2013, the moss garden at Duke University in Durham, North Carolina, features *Atrichum, Thuidium*, and *Bryoandersonia* species.

hornworts—collected by the team of rescuers.

The receipt of a substantial gift from a gracious donor enabled the mossy dream to become a reality. The Kathleen W. Smith Moss Garden was dedicated in May 2013. With a jigsaw puzzle of hundreds of harvested divisions planted tightly together, the new moss garden is thriving. Twice each day pop-up irrigation heads automatically provide a light and gentle watering for about eight minutes. Before this supplemental watering became routine, the mosses performed adequately well, but with the additional moisture they are booming. Stone water basins, boulders, and cobblestone paths have quickly gained a moss patina. Jones believes watering is key to achieving the best success (I heartily agree) and reports no negative effects from the use of city water.

To keep the garden looking fine, staff and garden volunteers provide daily attention to weeding and to repairing damage from squirrels and birds. In sections of the older moss garden, controlling obnoxious nutsedge (*Cyperus* species) and pervasive pennywort (*Hydrocotyle* species) continues to be challenging. Birds and squirrels toss moss fragments around with disrespect, so repairs are a consistent task. Typically, *Dicranum* colonies fall victim to this damage more than other species. While making repairs, workers collect fragments and scatter them around the fringe areas of the garden. And guess what? Yes indeedy, they too are spreading thanks to the consistent moisture of routine watering.

More garden expansion is on the horizon at Duke. Renowned Japanese garden expert Sadafumi Uchiyama, curator of the Portland Japanese Garden and third-generation gardener, has designed a new Japanese-style garden, and mosses are definitely on the plant list. Similarly, as part of a shift from a generalized approach to development of gardens with specific cultural themes, design of a Chinese garden is now under way. As the Asiatic gardens take on a new flair with ancient roots, the tradition of serene mossy retreats continues.

The Japanese Garden at the Missouri Botanical Garden, St. Louis, Missouri

"Garden of pure, clear harmony and peace" is the English translation of the Japanese term *Seiwa-en*, the official name for the Japanese Garden at the Missouri Botanical Garden in St. Louis. Created in 1977, this 14-acre garden reflects the influence of Shinto and Zen Buddhism, with an emphasis on nature. A 4-acre lake with undulating edges and circuitous stone paths provides the centerpiece of this stroll garden. Around every curve and across each elegantly arched bridge, vantage points offer glimpses of waterfalls, islands, trees, rocks—symbolic elements of traditional Japanese style. In the dry garden, carefully manicured white pebbles raked in patterns that characterize ocean waves surround a triad of boulders representing mountain islands.

As in many Japanese-style gardens, cherry trees dominate the landscape in the spring and maples dot the monochromatic landscape with pops of color in the fall. When blanketed in snow, the architectural elements and natural forms are distinctively impressive, particularly at night when the flickering glow from archetypal stone lanterns creates dancing patterns across the sparkling crystalline surface. Strategically placed benches provide opportunities to reflect upon the immense beauty of the surroundings and to take time for inner reflection.

Designed by Koichi Kawana, the garden has changed little over the years and authenticity has continued to be a priority. Kawana's conceptual plan and recommended plant list did not cite mosses. But determined indigenous bryophytes have moved into this Midwest garden, and the staff has welcomed their presence. When a patch of euonymus received the official designation as an invasive plant and was removed, gardeners discovered mosses living under it and encouraged their growth.

Raking white pebbles or sand emulates ocean waves. Stone columns represent islands.

With consideration for expanding the moss footprint, a small experimental plot is being developed. The elaborate planting method involves several layers, including plastic sheeting as a base, felt as a water-retaining substrate, a combination of common moss species (*Dicranum*, *Leucobryum*, *Polytrichum*, *Thuidium*, and *Hypnum* along with a *Cladonia* lichen), and a topping of tulle (fine net fabric). Some species are doing okay while others have not taken off in this 3-by-5-foot parcel, according to horticultural supervisor Ben Chu. Continued experimentation is planned with specific species placement, alternate planting methods, and supplemental watering routines.

A great garden destination throughout the year, Seiwa-en captures the intended essence of harmony and peace. Briefly, during Labor Day weekend each year, the serenity and calm are replaced with the melodic strains of Asian musical instruments and thundering booms of taiko drums as colorfully dressed celebrants gather for the annual Japanese Festival.

Humes Japanese Stroll Garden, Mill Neck, New York

As you ascend the steep steps of the hillside terrain, moss-covered rocks and stone lanterns adorned with moss facades lead the way to a hidden retreat reminiscent of a remote village in ancient Japan. In reality, you are on the other side of the world in Mill Neck, New York. At the John P. Humes Japanese Stroll Garden, trickling streams and cascading waterfalls trigger a sense of relaxation as you slow your pace to soak in the entirety of this calming approach. A gravel path symbolizes the flow of rivers to a crystal lake that represents the ocean.

When you arrive at the top of the hill, you experience the sensation of stepping back into another time and place. A reconstructed teahouse is a central focal point. Brought to this country in 1960 by the U.S. ambassador to Japan, John P. Humes, it exemplifies architectural design from the Ashikaga period (1477–1560). In the 1980s, this private residence was opened to the public. The mosses have had more than fifty years to form a gorgeous green expanse, enhancing the meditative charm of this notable Japanese strolling garden.

Climbing the steps toward enlightenment is a design concept exemplified in the John P. Humes Japanese Stroll Garden.

BOTTOM This moss lawn at the Humes garden exhibits a seasonal color change. Golden tones will shift back to green.

Ongoing maintenance—supplemental watering, weeding, and litter removal—has ensured long-term success of the Silvermont Moss Circle Garden.

Woodland Garden at Silvermont Park, Brevard, North Carolina

Opportunities to enjoy mosses exist in public gardens in small towns just as much as in urban areas. I was involved in the creation of one such garden, the Woodland Garden at Silvermont Park in Brevard, North Carolina, featuring a 160-square-foot circle of indigenous mosses as its centerpiece. It represents a model for community involvement in rescuing mosses from destruction at a nearby residential community and relocating them in a public park. Master Gardeners in Transylvania County, North Carolina, have been nurturing and tending this moss feature since it was installed in 2010 in hot temperatures under scorching sun. It continues to thrive, providing year-round pleasure. I feel like a proud mama and certainly appreciate the continued dedication of the weekly volunteers who keep it looking good with regular weeding and consistent watering during spring, summer, and fall months.

Master Gardener volunteers create a community moss garden in Brevard's Silvermont Park in June 2010.

Leila Barnes Cheatham Learning Moss Garden, Highlands, North Carolina

Over the mountain and through the snowy woods, I created a learning moss garden in the dead of winter (January 2012) at the Highlands Biological Station, a research facility connected with Western Carolina University and a national site for continuing education for students and professors from around the country. Located in Highlands, North Carolina, the Leila Barnes Cheatham Learning Moss Garden is nestled in a rhododendron alcove among fantastic ferns that occupy this ravine location. It features a number of native mosses in a planting design that emphasizes the beauty of the natural surroundings.

Furthermore, this learning moss garden continues the Highlands Biological Station tradition of providing identification labels for moss species along with other flowers, plants, and trees. It is one of the few places where you will find moss species identified with plant markers. Interpretive signage supports the educational mission of this research facility. During your visit, explore the trails that wind past a pond bog and lead you through old-growth forests with trees more than five hundred years old.

Leila Barnes Cheatham
LEARNING MOSS GARDEN

Explore the fascinating, miniature world of the 450 million-year-old Bryophyta family of mosses. In this serene retreat, observe the variety of textures, subtle shades of green and accents of sporophytic colors.

Bryophytes reproduce through a two-stage cycle:

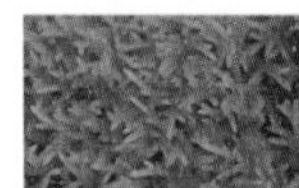

Gametophytic Stage
Plants produce egg and sperm for fertilization. Then in the 2nd stage...

Sporophytic Stage
Spores mature in capsules which spread via the wind to create new plants.

Also, bryophytes can grow asexually from leaf fragments.

Botanical characteristics of non-vascular mosses:

- NO roots - only rhizoids to attach to surfaces
- NO flowers - NO seeds
- NO cuticle - waxy leaf coating

Not all mosses grow in the shade ~ some species like sun exposures.

Moisture and nutrients are absorbed quickly through leaves only one-cell layer thick. Phenolic compounds enable mosses to tolerate extreme cold and to resist pests and diseases. NO fertilizers ~ NO herbicides ~ NO pesticides are needed for cultivating mosses. Water and walk on your mosses to encourage growth.

Bryophytes offer environmental solutions for:
water conservation • erosion control • stormwater filtration • nutrient poor soil.
Feature eco-friendly mosses in your own garden, moss lawns, stone patios and green roofs
to enjoy year-round green.

An interpretive sign at the Leila Barnes Cheatham Learning Moss Garden supports the educational mission of this garden.

Of special interest to moss enthusiasts is the week-long bryophyte workshop offered by Highlands Biological Station every other year. Academic credit is available for students completing this intense course of scientific study focusing on the botanical aspects of mosses and liverworts. The class is open to serious moss lovers as well. Experiential learning is emphasized through daily field trips and microscopic examination of bryophytes in the laboratory. If not so scientifically inclined, the general public and garden club members can learn about mosses through my periodic lectures and workshops where participants have the opportunity to make their own moss dish garden or live herbarium collection.

PRIVATE MOSS GARDENS

Hidden in cozy corners and tucked behind garden walls, private moss gardens range in size from intimate spaces in backyards to featured focal areas on expansive estates. Moss-laden paths meander through impressive displays of rhododendrons and azaleas, past filmy ferns and large-leafed hostas, leading to magical and mystical garden adventures. Some moss gardeners, inspired by travels, have succeeded in achieving the lofty goal of creating their own Japanese tea garden experience. Others bring mosses into their gardens (or allow them to invade) with the intention of recreating a woodsy environment that reflects natural surroundings.

Whatever the design style or size of these moss gardens, each one can be as quintessential as the next. As I connect with moss lovers around America and throughout the world, I feel privileged and honored to visit their private gardens. From extravagant expanses to enchanting elfin nooks, each moss garden is magical to me. With a camaraderie of spirit and shared sense of pride, we eagerly exchange experiences and discuss lessons learned in our separate moss journeys. We each possess a desire to create landscapes that feature our

favorite plants. Sharing our successes and inviting friends, family, and even strangers into our special spaces is common.

I sincerely hope that I will eventually get to travel across the country and around the globe visiting all the moss gardens on my personal bucket list. I anticipate that my destinations will grow in number as I discover more secret moss gardens and connect with horticulturists who share my passionate attitudes about mosses. In my dream world, I do not want to be a casual visitor and admire from afar. Instead, it will be my personal mission to connect with the caretakers directly involved in daily tasks. I hope that by sharing with each other, we can continue to build a body of knowledge valuable to us all.

I proudly showcase some of my own moss-scape accomplishments here and have included profiles of other accomplished moss gardeners throughout this book. These folks are just a sampling of mossers. There are more of us around the country and the world than one might suppose.

Norie Burnet's moss garden in Richmond, Virginia, is an exquisite example of a premier American moss garden.

My own moss garden

In the decades since 1990, I have followed my heart and gradually transformed my yard from a typical suburban landscape into my own moss haven. The exquisite beauty of my diminutive landscape reflects my connection to nature with a balance of serenity and whimsy. The curb appeal of my mountain bungalow is dominated by thousands of itty-bitty plants representing some thirty to fifty bryophyte types. The wall-to-wall carpet of harmonized mosses offers me continual pleasure throughout every season. By integrating an assortment of moss species, I have ensured that my garden landscape is always vital as these exceptional plants go through cycles of growth and reproduction.

As a moss garden designer, I admire and respect the value of Japanese designs. Nonetheless, I do not feel the need to restrict myself to a single school of thought. I believe you can create an inviting and serene space following any good principles of garden design that incorporate function and beauty as essential components. You can venture into new territory by adding your own imprint and personal preferences. I have synthesized a blend of my mountain heritage and worldwide design influences with my passion for mosses to fulfill my own ideals of a natural sanctuary. Green mosses reign supreme as my primary plants of choice, but I have also incorporated other plants and trees I like. Over the years, I have emphasized natural settings through individual vignettes that have evolved and converged into a whole. Like other Westerners, I am fond of colorful flowers that serve as occasional accents and add fragrant scents. Despite my partiality to indigenous plants, I am still drawn to perennials such as dancing daffodils (*Narcissus* species) and strong-scented stargazer lilies (*Lilium* species) native to other parts of the world.

While I did not feel the need to replicate a Japanese garden at my home, I have incorporated many elements that provide the same appeal. I planted a pyramidal-shaped stump

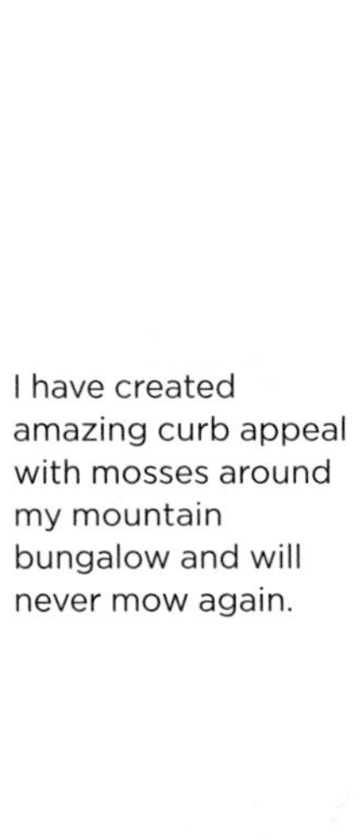

I have created amazing curb appeal with mosses around my mountain bungalow and will never mow again.

CLOCKWISE FROM LEFT
Japanese maples in my garden create a canopy of shade for mosses underneath and acknowledge a distinctive Asian horticultural choice.

With an eclectic style, my garden features unique logs and, of course, plenty of mosses.

This bonsai creation withstands all weather conditions to emphasize the miniature aspects of my moss yard.

found in the woods with a dwarf juniper and tucked decorative mosses and lichens into niches for my own interpretation of bonsai art. It sits near its natural counterpart—a moss-covered remnant of one of the mighty American chestnut trees (*Castanea dentata*) that once dominated forests along the East Coast of North America before being wiped out by a blight introduced in the early 1900s. The hemlock seedling (*Tsuga canadensis*) sprouting from this decaying chestnut stump has burst forth with new growth, requiring regular pruning to maintain small stature and spreading horizontal branching habit. The juniper bonsai tree has maintained its small scale because of the restricted root system.

I have added several Japanese maples that will maintain a small stature. I just could not resist this distinctive Asian horticultural choice. The gentle branches and vivid leaves of specialty species such as *Acer palmatum* (Dissectum Group) 'Tamukeyama' create a canopy of welcome shade for mosses underneath and a cool spot for sitting a spell on gazing benches. These dwarf maples add charm and complement rhododendron (*Rhododendron maximum*) and mountain laurel (*Kalmia latifolia*) indigenous to my mountain region.

Hummocks of *Polytrichum commune* and *P. juniperinum* provide dimension, in combination with *Thuidium delicalatum*.

The seemingly homogenous green of my moss garden is really a multiplicity of hues, textures, and colony shapes and sizes. Some sublimely blend together while others stand in stark distinction from neighboring species. Upright-growing hummocks of *Polytrichum commune* and *P. juniperinum* are discernible even from a distance with their massive size and their respective kelly-green and juniper-teal tints accented by burnt umber tendrils of older growth. Fluffy, thick mounds of *Dicranum scoparium* and *Aulacomnium palustre* contrast dramatically in color, with the former a deep emerald green and the latter a precious shade of peridot. The tactile pleasure of these super-cushiony species eclipses their loveliness. They feel absolutely "moss-some" to my feet.

Just the opposite is true of the extensive growth of *Marchantia polymorpha* liverworts. I do not like walking on its rubbery-feeling leaves and tough, spiky sporophytes. However, its petite palm trees with ochre-yellow pockets of spores hiding beneath the leaflike tops are a spectacular sight. The domineering growth of this liverwort is starting to overpower other moss sections. Alas, this weedy liverwort's days may be numbered in my garden due to this aggressive spread and the fact that I really do not like its black, slimy appearance in winter and early spring months. Since I am so keen about green (all the time), I plan on redefining this flat lawn area with more of my favorite moss, *Climacium americanum*. I have been impressed with this moss's remarkable ability to attach quickly and to green new areas with stunning intensity. As this forest of treelike dwarfs matures, colonies will provide a range of greens and the tiny tops will evolve into the largest moss crowns found in North America.

My garden will always be in some state of transition. During the winter of 2009 to 2010, I transformed my asphalt driveway into a mossy retreat. Just this summer, removing an overhang of native rhododendron opened up about 75 square feet of new pathway where, under the influence of Japanese garden design, I placed black stepping-stones surrounded with bright green *Atrichum undulatum*. It is my intention to maintain dynamic tendencies and avoid any static viewpoint. All the while, I take pride in knowing that this transformation from a grass lawn, asphalt driveway, and typical landscape plants is the successful result of my own diligent efforts to introduce these enticing plants on purpose.

Walking on mosses is encouraged in the Kenilworth moss garden in Asheville, North Carolina, which features more than a mile of mossy trails.

An expansive moss lawn provides the setting for impressive outdoor sculptures.

Kenilworth moss garden, Asheville, North Carolina

Once I became enthralled with mosses and determined that I wanted to learn more about gardening with these green jewels, I was surprised and delighted to discover one of the best moss gardens in America existed in Kenilworth, a neighborhood of Asheville, North Carolina, where both my mother and I spent our respective childhood years. As a young girl frolicking in the community splash pool with my sister, Kay Kay, in the 1950s, I did not realize that a hidden moss garden of major proportions was being planted right down the street. I certainly had no idea of the significant impact this garden's splendor would have on me as an adult. Decades later, the first time I visited the Kenilworth moss garden, as it is fondly called, I was certain I had entered a moss lover's dream world. The overwhelming green permeated my being with its breathtaking beauty and phenomenal sense of presence.

In the Kenilworth moss garden, green is profoundly all encompassing. The divine *Dicranum* and *Hypnum* moss lawn spreads like a massive carpet of emerald gems sprinkled with citrine crystals. Trails that snake through glades of glorious flowers and past vast displays

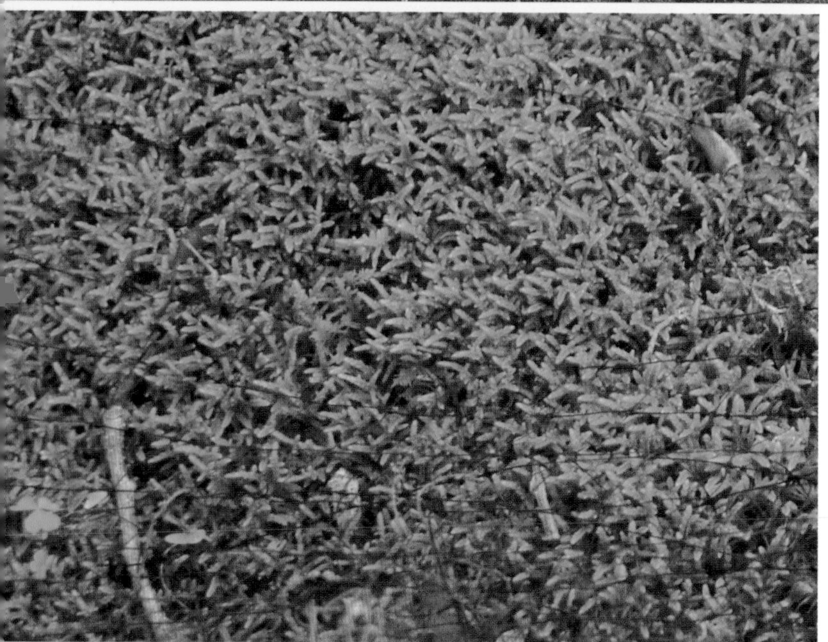

of azaleas lead to a series of secret destinations, each with its own special appeal. The original owner and landscape designer of this property, Doan Ogden, conceptualized green space as the universal backdrop for showcasing other plants, just as a graphic artist might conceptualize white space as essential to the whole. Today, the mosses serve as the platform for displaying stunning contemporary sculpture, Asian elements, and religious icons from medieval times. With distinctive panache, the eclectic style combines whimsy and dignity.

The current owners, John Cram and Matt Chambers, have kept this grand garden protected from the vicissitudes of time and elevated it to new heights. With the utmost respect for the conceptual design of Mr. Ogden, they assumed the role of privileged stewards with determination to raise this landscape from a state of benevolent neglect to its current status as one of the premier moss gardens in America. Maintaining more than 6 acres of land requires considerable attention and lots of hard work, but the expansive moss lawn and extensive trail system are proof that moss gardens can last for decades if properly maintained with loving care.

Even with a dedicated landscape crew, weeds can be overwhelming. To address the wearisome task of weed removal, John swears by a solution of one part Roundup diluted in four parts water. When talking about weeds, he references the real estate adage—location, location, location. If any aggressive plant or stray wildflower decides to locate in his mosses, it is considered a weed by his standards. Weeding is key, and he fanatically keeps up with the challenge of trying to nip the noxious plants in their early stages.

He has discovered that the use of wildlife netting helps reduce maintenance chores related to leaf litter and periodic droppings of male flowers or tassels from his huge oak trees. In the spring, these oak catkins can cover mosses like a brown blanket, but catching them on the netting makes removal much easier. Netting is beneficial in preventing the damaging effects of birds and critters, too. Power leaf blowers are essential tools as well. "The consistent use of blowers along the trails has allowed the moss footprint to extend further and further into perimeter areas, perhaps double, maybe triple," according to Cram.

The meditative aspects of this historic hidden retreat have a calming effect upon all who enter. As John puts it, "Its peacefulness, its chartreuse greens in winter when everything else is brown, contribute to the feeling of 'Zen-ness' and tranquility." This garden has lifted my spirits as much on repeat visits as it did on my initial otherworldly visit. While my moss garden offers its own delights, it pales in comparison to the acreage of this elite moss sanctuary. I could wander through this shady refuge for hours and hours. For others, and myself, I am grateful that my moss friends generously share their special place during garden tours held each June when it is open for public visitation. On those tours, with a twinkle in his eye and a slight smile on his face, John Cram beams with pride as garden visitors rave about his magnanimous moss garden.

Moss gardens at Willow Oak Farm, Madison, Georgia

Outside of Atlanta in the historic town of Madison, Georgia, I was honored to create a series of moss gardens for Ben Carter and his family on the site of a former cotton plantation. Ben wanted to create a unique space with American flair that incorporated native perennials, particularly mosses, and together we conceptualized a functional and serene retreat. With more than 2,500 square feet of mosses, these innovative gardens constitute one of the most ambitious undertakings ever attempted in contemporary moss landscaping. While other gardens may exceed the sum total of these gardens in size, I am confident that no other moss transformation of this magnitude has occurred within such a tight timeframe and provided such immediate gratification while promising to last for years.

In barely over a month from the initial site consultation, all the invasive privet had been removed from the property. Gigantic boulders and log benches were hauled in and carefully negotiated into position. Once the hardscape was prepped, we were ready to go. But a rare February snowstorm hit Georgia on the night the moss inventory was delivered from the North Carolina mountains to the site. Undaunted by the weather and without any worry about the effect of freezing temperatures, we just waited until roads were cleared and the mosses had thawed out a bit. And then, like magic, the entire moss landscape was installed within two days! Prevegetated moss mats rolled

OPPOSITE TOP An exquisite and expansive moss lawn of *Dicranum* and *Hypnum* species is the hallmark of the Kenilworth Moss Garden in Asheville, North Carolina.

Netting is used to protect the moss lawn from critters. It can barely be seen from a distance except around the edges.

A hollow stump at the entrance to Ben Carter's main moss room provides a portal into the miniature world of mosses.

out like carpets to facilitate an easy and fast installation. In this project, the substrate used for the mats is a weed barrier fabric that will help reduce any ongoing issues with persistent privet growth.

By mid-February, the dull winter landscape sparkled like a green oasis. When previously planted azaleas and rhododendrons burst into bloom, the mosses seemed to radiate from within. Native ferns and wildflowers installed a few weeks later further complement the mosses and add cohesion between sections. In subsequent plantings, more than a thousand perennials expanded the scope of the garden. Among the vascular plants is a rare azalea indigenous to the Great Smoky Mountains, *Rhododendron* 'Gregory Bald', which hybridized by itself in its natural setting.

With a variety of functional purposes, each moss terrace has its own distinctive whimsy and magic. In all moss gardens, barefootin' is encouraged to fully experience the tactile pleasures of the refreshing, damp mosses. One moss room is resplendent with thrones fit for any king or queen, fashioned from hollow tree stumps with soft, fluffy cushions of *Dicranum* moss. This intimate conversation area beckons visitors to relax over a game of chess or to sit and sip a fresh glass of lemonade. Large boulders provide additional seating for conversation among garden visitors. Moss carpets of *Atrichum*, *Plagiomnium*, *Rhodobryum*, and *Climacium* species cover the area. On a sunny hillside, a rainbow of sun-tolerant moss species offers additional focal interest.

The main gathering spot provides a relaxing view of the trout pond from seating along a huge log, rock boulders, or custom walnut benches. In the evening hours, a magical aura prevails as mosses glimmer in the light of dancing flames from the river rock fire pit. A portal view through a hollow log entices visitors to appreciate the petite aspects of mosses. In one corner, *Climacium* mosses rise

With privet removed and boulders in place, the site is ready for its mossy transformation.

TOP A rare snowstorm occurred on the mid-February night the mosses were delivered to Georgia.

Since mosses tolerate freezing temperatures, I was confident they could handle the snow.

TOP In a few days, the snow had melted and the mosses were fine and dandy.

Rolling out the green carpet, Mountain Moss crews create instant gratification with prevegetated moss mats.

above ground level like a miniature emerald forest of diminutive trees—a lush solution for this consistently soggy spot. The neon green of *Atrichum* species' new spring growth contrasts with the yellow-green tints of *Thuidium* species and beckons visitors to enter. The nuances of green and the variation in textures highlight the distinctive differences among moss species.

A moss-and-stone patio affords yet another vantage point to view trout swimming in the pond. To resolve erosion concerns and soften the harshness of the gravel surrounding the pond, colossal colonies of *Polytrichum* mosses line the cooling pool along the patio perimeter. These mosses offer a far more pleasing option than hay bales or riprap. With specialty mosses growing in the cracks between stones, the new rock wall and patio look like they have been there for years. Mosses add an instantaneous sense of permanence and antiquity to hardscapes.

The installed irrigation system keeps the mosses happy, and all species are thriving. Unfortunately, weeds have thrived, too. The first season required several major weeding sessions. *Climacium* moss has served well as the premier solution for naturally soggy areas and rainwater runoff paths. Several areas have required repairs due to the devastating antics of armadillos. As is the case with any moss garden, consistent watering, ongoing maintenance, and troubleshooting will be key to this garden's long-term success.

The throne room at Willow Oak Farm is fit for any king or queen with its emerald jewels. Flowering plants, shrubs, and trees further enhance this regal outdoor room where moss mats blended together later in the year.

LEFT Perched on the sitting log or on massive boulders, Ben Carter can relax with friends and family in this large outdoor room.

The moss-and-stone patio offers a view of the trout pond. Along the edge, colonies of *Polytrichum commune* hold soil in place, eliminating erosion issues.

bryophyte basics

BOTANY AND NATURAL HISTORY OF MOSSES

The plant kingdom consists of two basic types of plants: those with vascular tissue to transport fluid and nutrients, and those without. Mosses belong to the latter group of nonvascular plants, broadly known as bryophytes. As gardeners we customarily concern ourselves only with vascular plants, so entering the realm of moss gardening requires that we learn about how mosses are different from other plants.

Visions of green carpets enter our minds when we hear the word *moss*. Yet, a menagerie of misconceptions has hoodwinked many aspiring moss gardeners. First and foremost, lumping all moss species into one *moss* category creates confusion. Further, moss myths abound that muddle our efforts to garden with mosses. Our bewilderment continues with vascular plants that have the word *moss* in their common names. These quandaries hamper our attempts to understand how to grow mosses intentionally.

Mosses have been around for a long, long time—for many good reasons. Learning some botany will help you in building a sound foundation of practical knowledge that will lead to your own moss gardening success. This chapter explores the special features of mosses that influence moss gardening decisions and methods.

PAGE 70 The curly leaves are apparent on *Hypnum curvifolium* when you take a close look.

MOSS NOMENCLATURE

Distinguishing differences among moss species and how they vary in comparison to other plants is key to understanding how to use mosses successfully in planned landscapes. It's also important for gardeners to recognize that there are hundreds, even thousands, of separate moss plants in a single mound or mat of moss. Somehow it has become commonplace for people to refer to colonies as well as all bryophyte types in the singular—*moss*. The term *moss* implies homogeneity rather than reflecting the expansive number of separate moss plants that make up colonies or the vast variety of moss species. I, myself, have been guilty of this interchangeable use of the singular term to imply the plural. However, for horticulturists, botanists, environmentalists, or plain ol' moss lovers, it becomes important to move beyond grouping all mosses together as one.

Fortunately, a system of nomenclature exists to help us distinguish moss species from each other. Carolus Linnaeus is credited with establishing the first system for naming organisms (binomial nomenclature, or two scientific names) in 1753, but his knowledge of mosses was minimal. The man with the plan in the moss world was Johann Hedwig, who provided

Distinctive sporophytes mark the second reproductive phase of a moss's life cycle.

Hedwigia ciliata is named after Johann Hedwig, who developed nomenclature for bryophytes in the late 1700s.

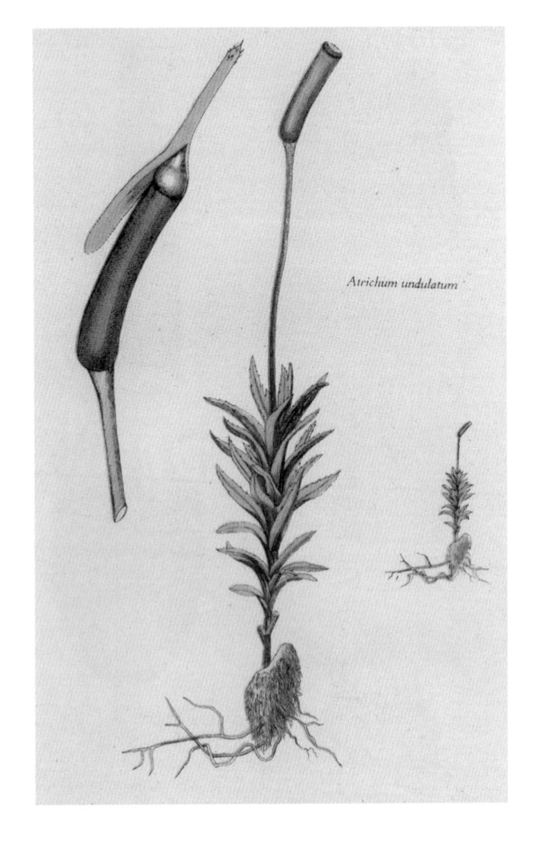

Like many early botanists, Johann Hedwig was a talented illustrator, as can be seen from his drawing of *Atrichum undulatum*.

us with names for all types of bryophytes. Hedwig, born in the Transylvania region of Romania in Eastern Europe, is noted for his attention to detail in his written descriptions and illustrations provided in his posthumously published book, *Species Muscorum Frondosorum* (1801, "Species of Leafy Mosses"). Other important scientific insights about mosses were published by German botanist Johann Jakob Dillenius in his *Historia Muscorum* (1741), which describes and illustrates more than six hundred species of true mosses, liverworts, lycopods, algae, lichens, and other lower plants. An American botanist, John Bartram, made this comment: "Before Dr. Dillenius gave me a hint of it, I took no particular notice of mosses but looked upon them as a cow looks at a pair of new barn doors." With this attitude and the previous absence of universal names and descriptions, it is not surprising that early botanists overlooked bryophytes and omitted them from their recorded observations and research.

Many gardeners shy away from scientific names and prefer to use common names instead. They just seem easier, don't they? However, common names can leave too much room for confusion. For one thing, they can vary from one geographic region to another. Sharing our knowledge or asking questions is more difficult if the terminology is vague or unfamiliar to other people. The acceptance of a universal way of naming plants enables us to communicate effectively with gardeners all around the world.

Aside from regional variations, common names can be misleading in other ways. For instance, many people use the term *pincushion moss* to refer to *Leucobryum* species. At least this common name reflects how this moss looks to the average person. But other quite different bryophyte species also form a cushion or mound shape (such as fluffy *Dicranum* or diminutive *Ceratodon* mosses). Thus, misperceptions are born. Certain common names do not make much sense except to bryologists or serious naturalists. For example, the common name for *Polytrichum* species is haircap moss, referring to tiny hairlike structures (calyptra) on top of the spore capsule. You might need more than 20/20 vision to see these teeny "hairs" with the naked eye.

From my perspective, if common names are used, they should reflect a particular botanical characteristic that helps the average person remember the specific moss. For instance, *Bartramia* species have spore capsules that in early stages resemble tiny green apples; hence the common name, apple moss. Of course, if the plant is not in its sporophytic stage, this name does not help in the identification of this particular moss.

Like me, you probably already use some scientific names for other flowers and plants like iris, forsythia, or hydrangea. If you are a beginning mosser, why not learn the real

Bryophyte bounty

The nonvascular plants or bryophytes include not only mosses (in the division Bryophyta) but also liverworts (Marchantiophyta) and hornworts (Anthocerotophyta), along with algae (Chlorophyta). At least ten thousand moss species, eight thousand liverwort species, and one hundred hornwort species exist in the world today. Some estimates range up to twenty-five thousand bryophyte species. Although mosses, liverworts, and hornworts share a unique two-stage reproductive life cycle, significant botanical characteristics distinguish the different bryophytes from one another.

It is simple to recognize a hornwort—if you ever see one, that is. With approximately one hundred hornwort species in comparison to thousands of mosses and liverworts, it is rare to find them. These plants grow tight to the ground in closely knit but flat colonies and have hornlike structures that protrude. The horns are part of their reproductive process where spores accumulate. The hornwort *Anthoceros carolinianus* growing in my yard is easy to spot because of its darker, denser green color and, of course, its horns.

The most visible difference between mosses and liverworts is in the leaves. Mosses have pointy leaves that grow in a spiral pattern around the main stem of the plant. Some mosses may trick you because their leaves appear round, but if you look closely with a loupe, you will usually see a tiny hair point on the tip end. The majority of moss genera have leaves with smooth edges (margins), although some have notched or toothed margins. In comparison, liverworts have round leaves that do not spiral. Most liverworts have a pattern of two leaves opposite each other with a third smaller leaf in between the two. This leaf arrangement continues along the entire liverwort plant stem, sometimes in an overlapping fashion.

Sometimes liverworts can be mistaken as creeping mosses from a distance. When you examine the individual plant, though, the leaf distinction is an obvious clue to whether it is a liverwort or a moss. To complicate the identification process, a few mosses, such as members of the genus *Homalia*, do have round leaves and no hair point on the tip.

Bryophytes include mosses, liverworts, and hornworts. Mosses (left) generally have pointy leaves, liverworts (middle) have rounded leaves, and hornworts (right) have horns where spores develop for dispersal.

Spectacular *Funaria* sporophytes arise from the charred wood of a fire. *Funaria hygrometrica* is a pioneer species, one of the very first plants to appear after a burn. Eventually other vascular plants overtake the area, and mature *Funaria* sporophytes become crimson and twisted together.

names for mosses instead of nicknames? With determination, I have learned formal names for bryophytes. It is not such a daunting task as one might imagine at first. Methodically learn a specific bryophyte by scientific name each day, week, or month until you build your own comfort level. Eventually, *Bartramia pomiformis* will be as easy to say as *apple moss*. Soon, these seemingly difficult terms will roll off your tongue.

BRYOPHYTE BEGINNINGS AND RANGE

Bryophytes have been around for 450 million years and were the first land plants on earth (although algae probably preceded mosses in the move from water to wetlands). They paved the way for other plants (tracheophytes) to grow into the lush green world we know today. About 50 million years after bryophytes appeared on land, other plants like ferns began to grow. It is mind blowing that mosses have been around for so long. Bryophytes watched the dinosaurs come and go. Living through all the changes in climate, mosses developed exceptional mechanisms for survival. Perhaps due to their simple structures, small size, and built-in chemical protection, mosses never needed to evolve like other land plants. Yet, scientists report some hybridization has happened over time.

As pioneer plants, bryophytes help break down rocks into soil to provide a nurturing environment for other plants to grow in. This role of bryophytes continues in modern times as an essential component of the forest life cycle. Mosses grow on decaying trees, offering a moist, protected place for seeds to germinate. After the destructive effects of a forest fire, the moss *Funaria hygrometrica* rises from the ashes as one of the first plants to return. The contrasting intense greens and brilliant ombré shades of sporophytes provide an impressive show against the ebony color of charred wood.

Even though mosses are minute in comparison to most plants, they represent a significant portion of the world's plant biomass. While there are more flowering plants on Planet Earth, bryophytes rank as the second largest group of plants, with between 15,000 and 25,000 species. Bryophytes occur on every continent in all types of environments and terrains, from lush tropical forests to arid deserts to arctic niches. Habitats range from the mountains to the sea. In my part of the world, the mountains of western North Carolina in the United States, more than 450 types of bryophytes thrive year-round. Bryophytes live all around us and survive well when deliberately introduced into gardens.

Mosses have connections across the globe. A moss species growing in your backyard might be found on another part of the continent or in a similar environment in a distant location across the ocean. Many mosses of eastern North America are also found in Europe or Asia. Bryologists Howard Crum and Lewis Anderson have traced the historical movement of mosses and say that 90 percent of the moss species found in the Ozark Mountains of Arkansas can also be found in other parts of the continent, while approximately 70 percent also live in the Old World. Species on the Florida peninsula are similar to those found in tropical lowlands. Mosses growing in the mountains of New York, New England, Virginia, Tennessee, and North Carolina are similar to European montane flora.

Some bryophytes live on land, growing on soil, rocks, decaying trees, and barks. Some epiphytic mosses hang from tree branches. While fewer in number, some bryophytes are aquatic (living under water). Bryophytes grow naturally in polar and alpine regions as well as tropical locations. Even in deserts, mosses have adapted to periods of desiccation with the internal ability to go dormant until moisture is once again available. Not all species are restricted to rural sites; certain mosses are cosmopolitan, living mainly in urban sites including the cracks of downtown sidewalks.

Ceratodon purpureus, a tiny, velvety moss, lives all over the planet. On the day that I was

Mosses skirt the bases of trees in nature. Can you tell if this tree is a result of Mother Nature or Mossin' Annie?

Mosses have sneaked into this tree square on the Georgia Tech campus in Atlanta. Wouldn't it be a good idea to plant these hardy mosses intentionally with the trees when greening cities?

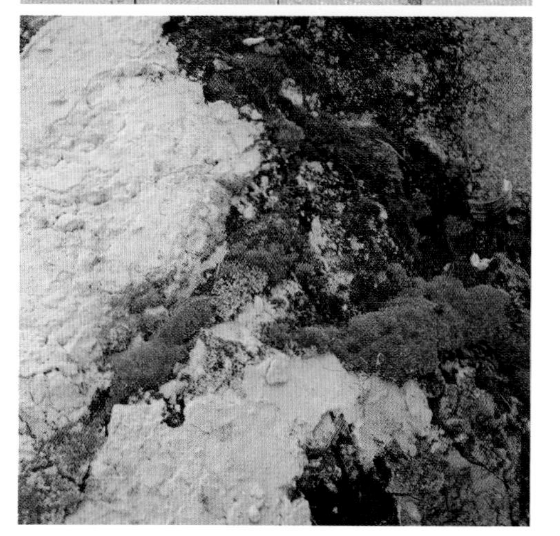

Tiny mosses appear in sidewalk cracks and around the edges of parking lots in urban areas. These *Entodon* and *Ceratodon* mosses are growing on the curb at the local recycle/trash center.

rescuing this urban moss species from the hot tin roof of an early twentieth-century house, a professor on an Internet forum was requesting photographs of the same moss to study during her expedition to Antarctica. I find it amazing that this moss lives not only in the temperate climate of the Blue Ridge Mountains but also in the extreme environment of the South Pole.

HOW BRYOPHYTES DIFFER FROM VASCULAR PLANTS

Bryophytes, like other familiar plants, require sunlight, carbon dioxide, mineral nutrients, and moisture to process their own food supply through photosynthesis. Beyond this basic similarity, bryophytes are unique in several key ways. Instead of relying on the structures and processes typical of vascular plants, mosses (as well as liverworts and hornworts) have distinctively different botanical features that allow them to not only survive but also thrive in a variety of geographic locations under various weather conditions and sun exposures. Bryophytes have no lignin (the chief component, along with cellulose, of the woody cell walls of plants), although they do have conducting cells to transport nutrients (so they are not exactly totally nonvascular after all). Special leaf features, rhizoids instead of roots, and sporophytes instead of flowers and seeds are additional major differences that set mosses apart from vascular plants.

To appreciate the minute details of mosses, a 10× close-up loupe is invaluable.

Note that some of the distinguishing characteristics of bryophytes can be viewed with your naked eye. Other details might require the use of a handheld magnifying glass or jeweler's loupe (10× magnification) for taking a closer look.

Absorption through leaves

Moss leaves are unique in that most have little or no cuticle, which is a waxy substance covering the leaves of other green plants. If you are familiar with rhododendron leaves, this waxy cuticle is obvious. The lack of a well-developed cuticle (or the presence of only a thin cuticle, as in the case of mosses like *Polytrichum* species) allows mosses to derive their moisture from rain, mist, and dew. Along with water, nutrients from dust particles are absorbed directly through the leaves of each separate moss plant. Nutrients also move with water in capillary spaces.

Thin skin and small leaves

Mosses hydrate quickly not only because they lack a cuticle but also because the majority of mosses have leaves that are only one cell layer thick. This means that mosses become saturated with moisture much faster than other plants. With such thin leaves, most have a translucent or see-through appearance. One notable exception to this general one-cell-layer rule is the leaf of *Leucobryum* species, which has multiple layers of cells. As the colony dries out, the top cell layer becomes opaque, presenting a whitish appearance—a good indicator in a moss garden that it is time to provide a drink of water.

Moss leaves always attach directly to the plant stem without a stalk. Each moss leaf has a thicker center section in the middle of the leaf called a midrib (nerve or costa). Individual

leaves are quite small, ranging from 0.02 inch to 0.12 inch long. Since the leaves are so tiny, novice moss gardeners mistakenly think the individual plants in a colony are the leaves, but in fact a colony is a group of moss plants, each with its own tiny leaves.

Rhizoids versus roots

Bryophytes have no root system. When you look at the underside of a moss colony, you see a mass of rootlike structures called rhizoids. Like roots, these filaments help the plants attach to surfaces. The threadlike rhizoids can find ways to attach in minute niches and cracks of rocks, soil, and even concrete. Mosses can even attach to vertical surfaces. As delicate as these rhizoids seem, they have the capability of holding mosses in place during high winds or heavy rains. Most important, rhizoids do not feed moss plants as roots feed other plants. No nutrients are transported through internal vascular systems. However, rhizoids do form an intertwined mesh that wicks water by capillary action, aiding in water retention.

No flowers, no seeds

Mosses never have flowers and therefore no seeds for reproduction. Instead, mosses reproduce in a two-stage cycle. The moss equivalent of a flower is the spore structure or sporophyte (the second stage of the reproductive cycle), and the spores within the capsules equate to seeds. Sporophytes offer brilliant colors and have a distinctive beauty that can be compared to a blossom's. When the spores have matured inside their petite capsules, the wind and rain serve as means for dispersing the spores to new locations. Before mosses achieve this second stage of reproduction, you may be surprised at what happens first. Hint: It starts with *s* and ends with *x*.

LEFT Certain mosses look dramatically different from wet to dry, making identification a challenge. *Hedwigia ciliata* is one species that exhibits major change.

Sporophytes could be considered the flowers of bryophytes. Capsules hold spores instead of seeds.

TOP Year-round green means winter glory. Once the snow melts, mosses reveal their green appeal again.

The translucency of leaves one cell layer thick is obvious in this example of *Rhodobryum ontariense*.

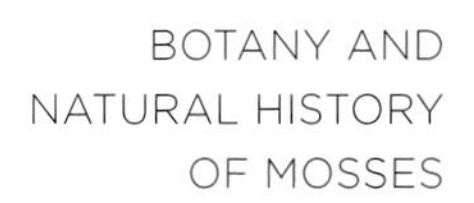

Resistance to freezing, predation, and disease

A plant with its own antifreeze? Yes, mosses have evolved to have their own internal protection against cold. Millions of years ago when mosses first appeared on our planet, they spent energy on developing internal biochemical mechanisms (phenolic compounds) for survival. These compounds help mosses to withstand subfreezing temperatures and also trigger dormancy as a survival technique when mosses' environment is too hot or too dry. Other phenolic compounds make mosses taste bad so that animals and insects do not find mosses appealing, while still others help to deter any diseases, as mentioned earlier.

Colony living

Mosses are social and live together in colonies. This close grouping of plants growing together is yet another way mosses retain moisture. Moss colonies do not necessarily contain only one kind of moss; various moss genera can live together harmoniously.

Mosses that grow upright are acrocarpous types. Those that grow sideways, prostrate to the ground (or substrate surface) are pleurocarpous mosses. Growth of acrocarpous mosses usually increases the loft or thickness of a colony, whereas pleurocarpous mosses display visible expansive, outward growth patterns.

It should be noted that certain mosses are more aggressive growers than others and may use a host moss colony as a preferred substrate. I learned this lesson firsthand one winter after having created an impressive spiral design using *Ceratodon* and *Leucobryum* species. When the winter snow melted after covering this impressive medallion for more than two months, I was dismayed to discover that *Thuidium* mosses had invaded my creation, compromising the original design. Eventually, *Atrichum* species moved in as well. If you want to maintain a static design concept, you may actually have to weed certain mosses out of other mosses.

TOP TO BOTTOM I rescued this huge *Dicranum scoparium* colony before the highway mowers came through. The snow disc used as a size reference is about 2.5 feet wide.

Pleurocarpous mosses have delicate-looking rhizoids that grow sideways and tenaciously hold mosses in place even during high winds.

Acrocarpous mosses grow upright. Check out this cross-section of *Atrichum angustatum*.

Bryophytes in brief

* Mosses, along with their cousins liverworts and hornworts, have the claim to fame as Planet Earth's first land plants. Classified as bryophytes by scientists, mosses are 450 million years old.
* Mosses live in all types of climates as long as moisture niches exist.
* Mosses have great potential to grow in all types of environments, especially if conducive conditions are present already or manipulated by gardeners to meet growth requirements.
* Without much of a cuticle, and having leaves that are only one cell layer thick, mosses absorb moisture quickly. Along with water, mosses acquire sustenance from dust particles directly through leaves.
* Rhizoids hold mosses in place but do not feed moss plants like the roots of vascular plants.
* Without flowers or seeds, mosses grow from spore dispersal, fragmentation, and gemmae vegetative reproduction.
* Mosses are impervious to cold temperatures, possessing their own built-in anti-freeze, and may exhibit signs of expansive growth or display reproductive sporophytes in winter months.
* Internal biochemical properties (phenolic compounds) protect mosses from all types of insect pests and herbivores as well as diseases.
* Mosses require moisture to thrive as plants; water is also essential to support reproductive processes.
* Moss plants (gametophytes) and reproductive structures (sporophytes) look different from each other and go through transitional stages.

HOW MOSSES REPRODUCE: THE BRYOPHYTE LIFE CYCLE

Mosses have an unorthodox reproductive life cycle in comparison to vascular plants, and this life cycle has significant implications for successful moss gardening. Most of us grasp the concept that flowers produce a seed that grows into a new plant. Remember, however, that mosses do not have flowers or seeds. So just how do mosses reproduce?

Mosses actually have three different methods of reproduction: via spores (two-stage sexual reproduction), via vegetative fragments (asexual reproduction), and via gemmae or propagules (vegetative reproduction, another form of asexual reproduction). Vegetative fragments and gemmae are a means of short-range dispersal via water. Spores do not weigh much and therefore can be transported easily by the wind, sometimes over long distances. Insects, birds, and animals can carry fragments and propagules to faraway places as well. My bryologist friend Janice credits her turtle with spreading moss fragments throughout her garden.

Spore dispersal (sexual reproduction)

Bryophytes reproduce through alternate generations that reflect the two main stages in their life cycle—gametophytes (green plants with, believe it or not, sexual organs) and sporophytes (spore-bearing plants). The first stage of reproduction is where the plant (gametophyte) produces eggs and sperm cells. Upon fertilization, spores mature in the spore capsule (urn) in the alternate generation (sporophyte). Of interest to botanists, the first generation (gametophytes or plants) has only one set of chromosomes (haploid), while in the second stage, the sporophytes have both X and Y chromosomes (diploid).

Mosses actually have sex. The male organ (microscopic in size), named the antheridium, releases the male sperm. They swim across the moist plant surface in an effort to reach the female structure called the archegonium (also microscopic), which houses a single egg. Johann Hedwig, the eighteenth-century botanist who gave us scientific names for bryophytes, is credited with the discovery of these microscopic sex parts. Before he identified these reproductive organs, mosses were thought to reproduce in other ways. His astute observations and the development of better microscopes factored into this major scientific breakthrough in the study of bryophytes.

Acrocarpous (upright-growing) mosses typically have spore structures that arise from the top of the plant. The sporophytes of pleurocarpous (sideways-growing) mosses grow from the main stem or branch. There are usually five egg receptacles on each branch of pleurocarp species, while acrocarps have about five per plant. Water is important to the fertilization process because it allows the flagella propelling the sperm to splash in a swimming motion toward the egg.

The alternate or second generation of a moss plant begins as the fertilized egg develops during the sporophytic stage of reproduction. A tiny stemlike structure called a seta (a sporophyte stalk) starts to develop. Sometimes the setae are short—less than 0.03 inch—or they could be as tall as 4 inches, depending on the bryophyte type. During this sporophytic stage, moss setae and spore capsules display an array of eye-catching colors—bronze, gold, scarlet, orange, and yellow, as well as a parade of green subtleties. In some mosses, the sporophytic stage may be apparent for months as they mature. The drama of sporophytes in transition is particularly long lasting for *Polytrichum commune*. For at least six months, the sporophytes put on a show, starting with the earliest spikes, proceeding through color bursts, and finally ending with brown shriveled capsules. In contrast, liverworts are far less showy, with clear or whitish setae and sporophytes that may last only a few days. A liverwort exception: *Marchantia* spore structures are present for long months and look different from other typical sporophytes.

Multiple sporophytes grow on different branches of pleurocarpous mosses, while usually a single sporophyte emerges from the top of an acrocarpous moss plant; only a few mosses (*Climacium* and *Rhodobryum* species) have multiple sporophytes originating from the same spot. When sporophytes start to emerge, you will need to look from different angles to see them. If you rub your hand lightly across the moss colony, you can feel emerging pointy spikes. A wonderful moss garden reward occurs when raindrops encapsulate the sporophytes and the sun glimmers on your mosses. You will experience a captivating beauty equal to any flower garden.

This is definitely a time when a hand lens (loupe) is essential for taking that closer look. Go ahead—be a Peeping Tom (or a Peeping Annie as the case may be) and take a sneak peek at the sporophytes. Each species has a capsule with distinguishable features during this sporophytic stage. Young sporophytes grow upright and typically stand above the moss plant. Certain mosses like *Diphyscium* have urns at the base of the plant. Sometimes sporophytes are hidden among the gametophytes of the moss colony. I will never forget the day when I first found the sporophytes of *Hedwigia* buried deeply among plants in the colony. Weeding can have the advantage that it provides this intimate observation of minuscule details.

As the spores mature in the capsule, it swells and takes on a particular shape. Some spore capsules are cylindrical, some are spherical, and others may have almond-shaped heads that nod down. Observing sporophytes and the

Sporophytes are the second stage of reproduction for bryophytes.

OPPOSITE Acrocarpous (upright-growing) mosses like this *Climacium* species typically have spore structures that arise from the top.

Sporophytes enhance the beauty of mosses. Different species display them at different times of the year, providing all seasons with these extra delights.

OPPOSITE
In this liverwort, female *Marchantia* spore structures (antheridophores) look like palm trees and male spore structures (archegoniophores) look like flat umbrellas as they emerge and mature.

shape/height of the capsules aids in the correct identification of a moss species. The covering that initially protected the eggs in the archegonium expands after fertilization to become the calyptra (which to me looks like a flying nun hat) that eventually blows away from the sporophyte to reveal the bloated capsule. The operculum, which looks like a pointy top hat, is the covering over the top end (mouth and teeth) of a spore capsule. When it falls off, spores are ready for release. Usually mosses have a round opening, but there are exceptions (*Andreaea* mosses have slits instead). Around the edges of the opening, tiny peristome teeth help to propel the spores out to new locations. While some mosses have eight or sixteen peristome teeth that move in response to changes in humidity, it is not uncommon to find pleurocarpous mosses with thirty-two peristome teeth around the mouth (just like humans).

As the spore capsule (urn) expands, meiosis (cell division during which the number of sets of chromosomes in the cell is reduced to half the original number) occurs, facilitating the growth of spores. The number of spores in a single capsule varies. *Dawsonia* takes first prize

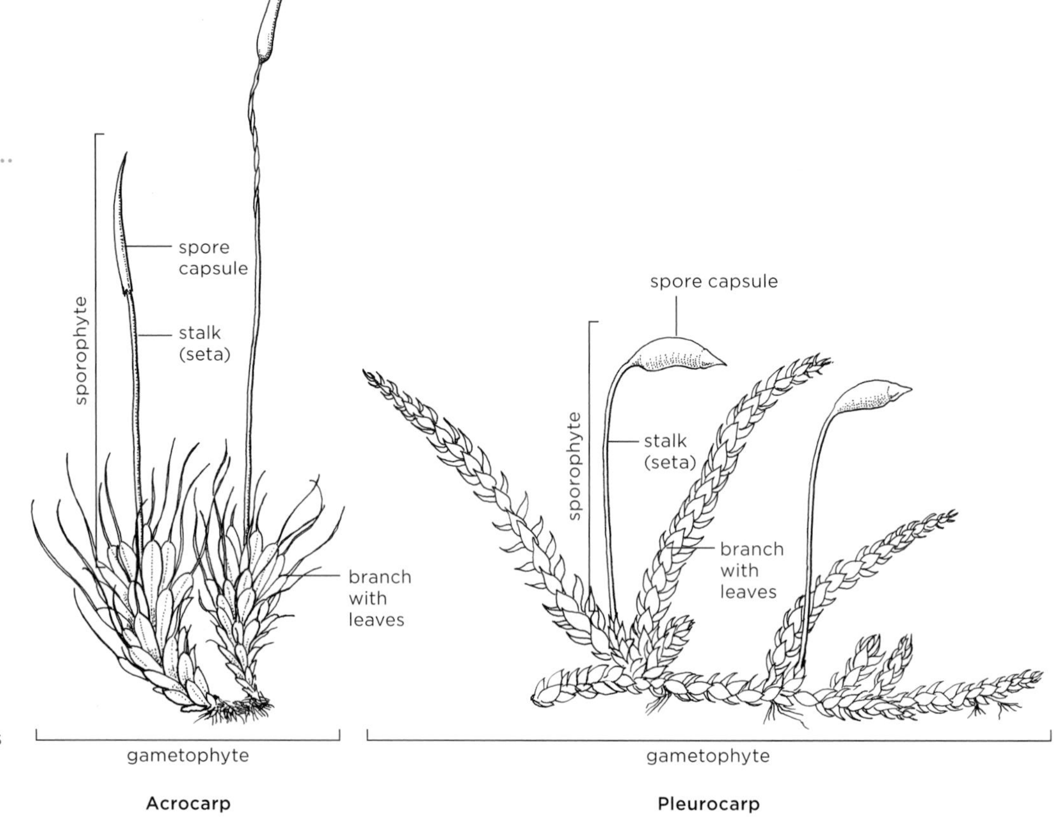

Differences in the basic structure of acrocarpous and pleurocarpous mosses

with the most—up to eighty million spores encapsulated. *Polytrichum* takes second place in the world but number one in North America—a mere million spores. Pause and think about these staggering numbers. Since the capsule is smaller than a grain of rice, the spores are infinitesimal in size.

When spores are mature and ready for dispersal, natural forces—primarily wind and rain—help distribute spores. A visiting bee or pesky squirrel can trigger sporophytic bursts. Likewise, if you barely touch the tops, a cloud of spores could float around you. Usually the release of spores occurs when the plants are dry. The spores can be released all at once or in smaller bursts prompted by humidity levels. While some mosses release spores in little dabs like a saltshaker, others exhibit blasts that are more intense. The power of spores exploding can be extraordinary. Vortex rings occur during *Sphagnum* mosses' spore dispersal process; the spores disperse at speeds of 65 miles per hour in a mushroom cloud that resembles a nuclear explosion.

My favorite interactive spore experience occurred while I was sitting on a roof rescuing *Entodon* mosses. It was a spectacular fall day with Carolina blue skies, a slight, wispy breeze, and a warm afternoon sun. I had faced my fear of heights and climbed over that formidable top rung of the ladder to walk gingerly around the roof. While reveling in the beauty of my surroundings, I was excited to see at least 75 square feet of *Entodon seductrix* densely covered with sporophytes. Not exactly sure-footed on a roof, I tend to sit down and scoot around, dragging my

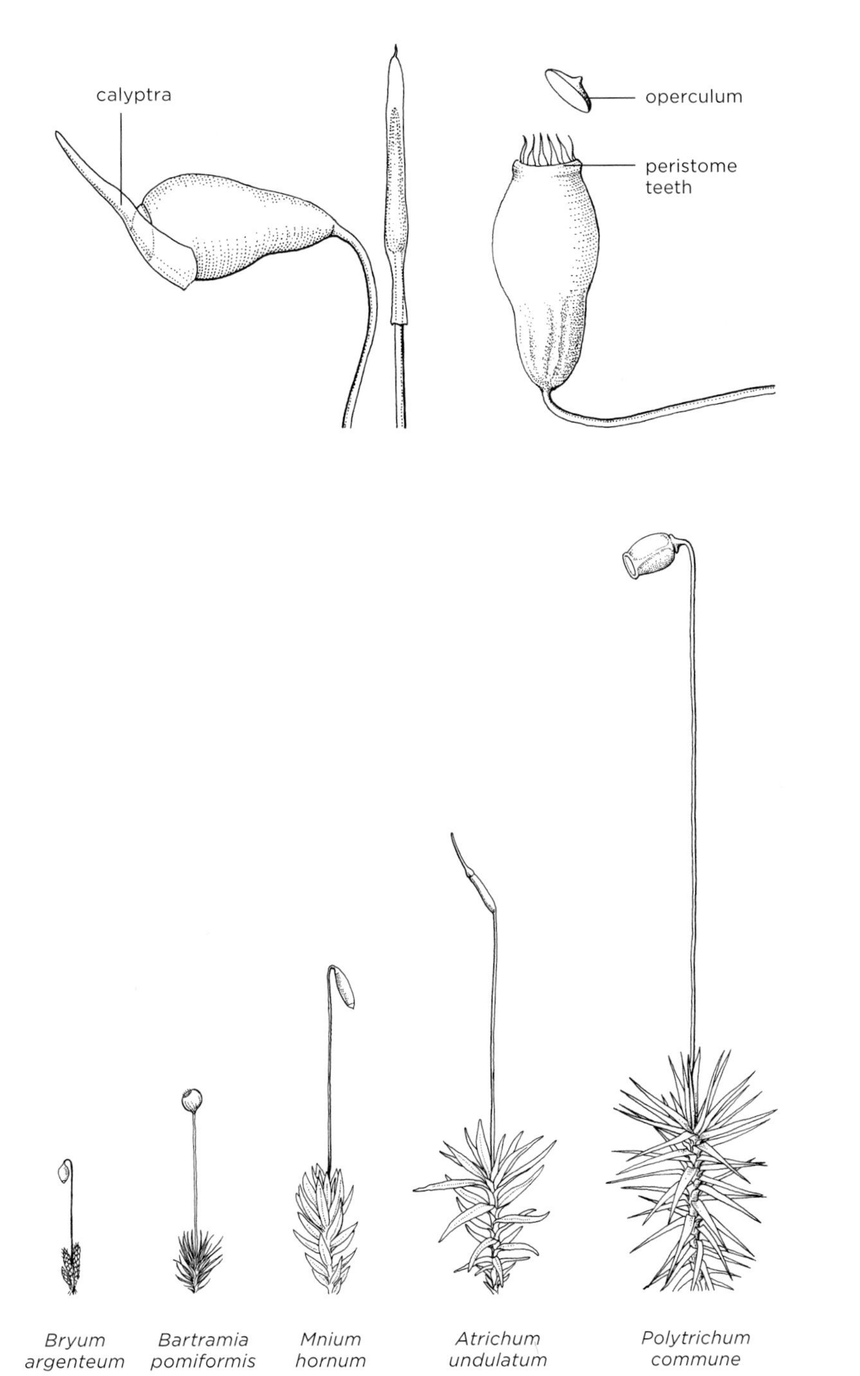

Parts of the spore capsule

Examples of differences among species in sporophyte stalk height and capsule shape

moss sled with me. As I reached to remove the first colony, a yellow haze erupted. I let out a squeal of delight and startled my spotter on the ground. Every time I barely touched the sporophytes, more golden clouds appeared. As I marveled at this stellar display, I felt like Mossin' Annie in Wonderland. Then my next thought was that I hoped my moss spore happening would not affect me the way the mushrooms did Alice in Lewis Carroll's book. I am happy to report that I had no adverse reactions, although my immense fascination was more than a bit magical.

For most species, this sporophytic event occurs annually, but how long they continue to spore is dependent upon climatic factors. The life cycle of a sporophyte might be days, weeks, months, or more than a year. Since mosses do not necessarily follow typical seasons of other plants, sporophytes may appear on colonies in spring, summer, fall, and even winter. Do not expect all mosses to have displays within the same timeframe. You can expect colonies of the same species in close proximity subjected to the same environmental conditions to evidence both gametophytes (stage one) and sporophytes (stage two) at the same time. Sometimes this reproductive process seems to occur once a year, yet in certain species sporophytes occur several times a year.

Spores can grow into new plants if they land in a suitable place in which to live and thrive. When a moss first grows from a spore, it starts in the protonema stage with hairlike filaments forming a tangled green mat. Eventually, it will grow into a plant with a stem and leaves. These individual plants grow in colonies, which may include other bryophyte types, ferns, fungi, or vascular plants. Plant colonies have both male (antheridia) and female (archegonia) reproductive organs growing on the same plant (if it is monoicous or bisexual) or nearby as separate plants (if plants are dioicous or unisexual). Separate male and female plants (dioicous) are the primitive form. Mosses that are monoicous (males/females together) seem to be evidence of hybridization and polyploidy (multiple sets of chromosomes). Genetic research cites that chromosome numbers indicate some hybrids have resulted in new species and dwarf males.

Under moist conditions, the sperm produced by the male organ swims to fertilize the egg produced by the female organ, and the cycle begins all over again. How effective are spores in establishing new plants and spreading to new areas in your garden? It is probably the most challenging way to introduce mosses. I've not really experimented with this planting method and have not found any substantive answers in documented research. Based on my own observations, *Polytrichum* is effective at spore dispersal. Unexpectedly, I have seen this hardy moss species appear in areas of my moss garden where I did not intentionally plant it, and it is doubtful that other vegetative growth methods took place due to the distance of originating colonies.

After the sporophytic stage, moss plants supporting these reproductive structures may go into a dormant stage or die back. The green color of the plants may change to another duller color, brownish or brick-colored shades. Do not worry. This is just a stage. Within weeks, new intense green growth appears in these sections of the colony. Fresh growth occurs from the tips of plant stems as well as stem branches. It is important to recognize transitions as a normal part of the life cycle of mosses. I hope you, like me, will find joy in all stages of growth throughout all seasons.

Fragmentation (asexual reproduction)

Imagine being able to propagate moss by tearing apart a colony, pulverizing or cutting the plants into small pieces, and then distributing them across your landscape. These hardy little plants are determined to spread one way or another. Fragments of plants can grow into new

White leaf tips break off *Leucobryum glaucum* colonies during asexual reproduction.

plants when moved by wind, water, or birds or other animals disturbing their habitats. These fragments don't even have to land right side up since mosses can grow from the base or stems of fragments. *Leucobryum* moss can grow into what I call a moss cookie with green growth on all sides. I have heard of giant moss balls that roll across the moors of Ireland.

Fragmentation (asexual reproduction), a vegetative reproductive process akin to cloning, is yet another survival method of these miniature plants that traces back to moss origins. The accidental breaking apart of plant parts by animals or from erosion occurs because zones of weakness exist that allow segments to break off easily to grow into new plants. An outstanding survival mechanism, fragmentation makes it possible for gardeners to establish expansive new areas of moss or make repairs to sections damaged by digging critters. Mosses tend to grow slowly in comparison to other plants, especially grasses, but using extensive fragmentation planting methods facilitates faster results.

Gemmae (vegetative reproduction)

That's not all, folks. Besides sexual reproduction and fragmentation, vegetative reproduction (another asexual means) can yield new moss plants. This process is a bit more complicated botanically. It involves the production of small cups, usually growing directly off stems or leaves. Many are visible only under a microscope. In these itty-bitty cups, tiny green gemmae, which get their name from the Latin for "jewels," may develop. Also known as propagules, these consist of a single cell or cluster of cells that separates from the parent plant to

Preserving natural ecosystems

With lingering childhood memories of making fairy gardens using sparkling mica rocks and soft, velvety mosses and with visions of magnificent gardens experienced during his world travels, Robert Balentine wanted to feature broad expanses of mosses to complement the variety of native plants in the gardens at the Southern Highlands Reserve. On top of Toxaway Mountain in North Carolina, along the Blue Ridge Escarpment (where mountains and steep cliffs abruptly abut the flatlands and rolling hills of the Piedmont), this 120-acre private reserve is dedicated to preserving and advocating the value of natural ecosystems through education, restoration, and research.

At 4,500 feet above sea level, the garden can be buffeted by brutal winds—sometimes up to 80 miles per hour. These winds stunt trees and bushes, but the mosses thrive due to positive factors such as abundant rainfall and high-elevation mists. Design professionals, local artisans, consultants, and staff made up the team that assisted Robert and his wife, Betty, in achieving their dream over a period of twenty years. You might assume that this successful Atlanta businessman (2013 recipient of a lifetime achievement award from the Atlanta Chamber of Commerce) and environmentalist (founder of the Southeastern Horticultural Society and TEDx lecturer) would sit back and just enjoy this incredible mountain haven. However, on his frequent escapes from the hustle and bustle of urban life, Robert relishes the time spent establishing and maintaining this impressive reserve. Like any other moss gardener, he considers pulling weeds and picking up litter to be routine tasks.

In a conscientious effort to plant mosses, locals helped gather *Hypnum* and *Thuidium* species from fringe areas of the property and relocated colonies to prominent locations along the driveway and near the research headquarters. Robert joined in the process of replanting these mosses in close proximity to attain a massive moss carpet. He recalls using twigs to hold down the edges of colonies planted right next to each other. Today the medley of mosses have thoroughly knitted together, showcasing delicate wildflowers and rare azalea species. Sometimes these fern mosses are green, and at other times, they have a golden hue. As for his favorite moss species, Robert particularly likes *Polytrichum* when its sporophytes are topped with their "flying nun hat" calyptras.

Robert Balentine may have more land than most of us, but his heart is right in line with moss gardeners like you and me. The serene aspects of mosses touched him as a child and continued to capture his interest as he visited gardens in America, Europe, and the Far East. Robert's travels have reinforced his love of mosses and have influenced his decisions regarding garden design at the Southern Highlands Reserve. In his own words, "It is fun to integrate the amazing textures, colors, and forms of mosses." Recognizing the soul-restorative aspects, Robert wishes he had "planted more mosses—and sooner."

ABOVE At the Southern Highlands Reserve, Robert and Betty Balentine have created a retreat and research center for the study of native plants.

LEFT Thriving *Hypnum imponens* and *Thuidium delicatulum* display their transitional golden color.

New growth will emerge from these male *Polytrichum formosum* cups.

Propagules are beginning to form on the inside of *Marchantia polymorphus* splash cups. Others have already splashed out.

form a new organism. The reality is that most mosses produce gemmae or other vegetative propagules, but these cool cups are not the rule. Gemmae can be dispersed to new locations and will grow into new gametophytes. Once again, water is a primary distributor. When rain hits the cups, the gemmae splash out for dispersal.

MOSS FAKERS

With an assortment of plants that are called mosses, it can be confusing to differentiate moss imposters from true mosses (bryophytes). If you have purchased a plant with a common name that includes *moss*, it may not be a true bryophyte. For instance, the Irish moss and Scotch moss sold in nurseries and garden centers are both flowering plants in the genus *Sagina*. Hint: If your "moss" plant comes in a pot with roots and it has flowers, you have purchased a vascular plant—not a true moss.

It is no wonder that gardeners are bewildered with the abundance of vascular plants, lichens, lichopods, and even algae having the word *moss* in their common names. Once again, I will reiterate that scientific names help us avoid misunderstandings. Misperceptions about these plants can result in frustrating results when they do not perform well in your garden. To avoid disappointment, let's become more knowledgeable about moss fakers.

Lichens

Cladonia rangiferina (reindeer moss, caribou moss, reindeer lichen) is naturally a light gray or greenish color. It is composed of intertwining branches that frequently grow into mound shapes ranging in size from a tennis ball to a basketball. It grows in hot and cold climates alike. Since it is extremely cold hardy, reindeer moss can be found in alpine tundra areas around the globe. Reindeer and caribou eat this lichen as a means of resisting cold temperatures—hence the common name reindeer moss. Yet, cladonia lives in warmer climates as

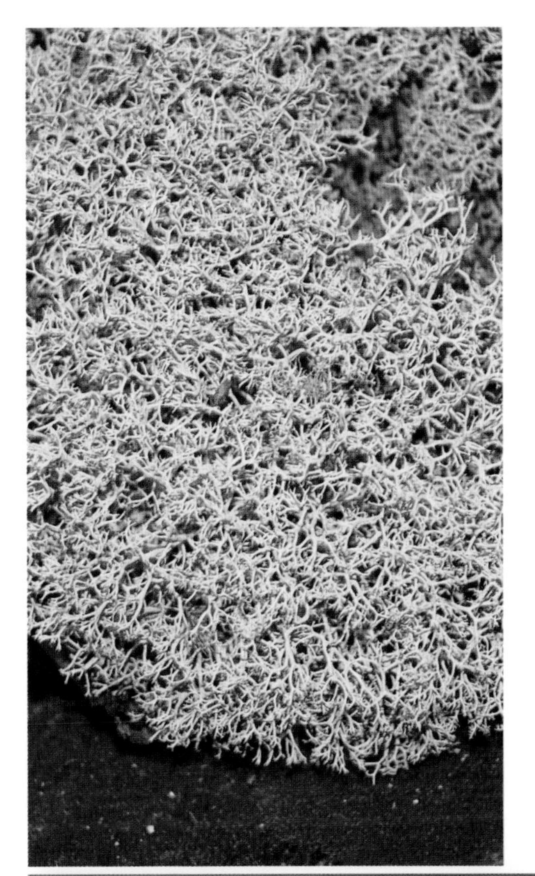

TOP Commonly called reindeer moss, this plant is actually a lichen, *Cladonia rangiferina*.

Cladonia cristatella lichens have brilliant red tops. They are known as British solders in the U.S. and as matchsticks in the UK.

LEFT Spanish moss is not a moss but instead a bromeliad.

RIGHT Don't be fooled by Irish moss or Scotch moss sold in a pot at your local garden center. They are vascular plants with roots, flowers, and seeds.

well (such as Florida) and grows for miles along open roadsides in sections of the Carolinas.

While cladonia is a threatened plant in some places as cited in the United Kingdom Biodiversity Action Plan, vast amounts continue to be harvested throughout the world. In the United States, reindeer moss is sold in mass quantities to florists. It is dyed chartreuse green and a substance is added to make it retain its spongy texture. Interior designers and floral enthusiasts proudly use decorative bowls of these reindeer moss balls. They have effectively preserved these lichens for display and ensured they will never live again as plants. If you purchase these products, you may be supporting destruction of natural habitats and illegal harvesting methods.

If you want to use this lichen as a decorative element in your moss garden, it prefers to live in a sunny location on a hillside. It is important to include the original soil when you relocate cladonia. Although lichens like some moisture, they like to dry out, too.

Vascular plants

Tillandsia usneoides (Spanish moss) gracefully drapes from the branches of *Quercus virginiana* (southern live oak), *Taxodium distichum* (bald cypress), and other trees. Familiar sights in the southern United States, these whitish stringy plants hanging from limbs do not permanently damage the tree but they do slow down growth a bit. Spanish moss is not a moss. For that matter, it does not even grow in Spain except in greenhouses. To my surprise, it is a flowering plant in the bromeliad family related to pineapple plants. Landscape designers frequently incorporate spotlights at night to emphasize the drama of tillandsia adorning trees that are centuries old. Florists and crafters use Spanish moss as a filler or decorative dead element. When people associate chiggers with moss, they usually are referring to what they know as Spanish moss.

Sagina subulata (Irish moss, Scotch moss, miscellaneous *pearlwort* names including moss pearlwort)—a vascular plant native to Europe with roots, flowers, and seeds—can really fool you. It looks like a moss (*Dicranum scoparium*), but it does not grow like a true moss. In colder regions, this perennial ground cover dies back in the winter, in contrast with cold-tolerant bryophytes. Cultivated by wholesale nursery suppliers, Irish moss and Scotch moss are sold in garden centers and nurseries. Many gardeners express disappointment with the pot of moss they bought at the big box store when it does not grow like a moss should. Mosses get a bad rap when confused with *Sagina* plants.

Phlox subulata (moss phlox, moss pink, creeping phlox) is a favorite choice of rock garden aficionados and has a major colorful impact in early spring in my region. One of my own favorite memories is my grandmother's rock wall capped with pink, white, and magenta phlox flowers. It can be an effective spreader as a ground cover, too. It can give a similar visual experience to moss when it is green during summer months. Alas, another vivid memory of phlox is that it is prickly to sit upon—not nearly as inviting as mosses in this respect.

Algae

Aegagropila linnaei (cladaphora balls, marimo balls, lake goblins, moss balls) is not a moss but a species of filamentous green algae. Velvety green algae balls grow in Northern Hemisphere lakes in countries such as Japan, Iceland, Scotland, and Estonia. While Anton E. Sauter is credited with the discovery of these algae balls in Lake Zeller, Austria, in the 1820s, we can thank the Japanese botanist Tatsuhiko Kawakami for giving us the common name marimo balls in 1898. In Japanese, *mari* means bouncy play ball and *mo* is a reference to plants that grow in water. *Aegagropila linnaei* has been a protected species in Japan since the 1920s, with cultural celebrations held annually at Lake Akan, Hokkaido. In Iceland at Lake Mývatn, the ball-shaped alga was given protected status in 2006. As with cladonia (reindeer moss), I suspect abusive practices in the collection of *Aegagropila linnaei*. It is a popular addition to aquariums and sold extensively on the Internet. I get proposals every week from cladophora suppliers (mainly in Japan and Hungary) wanting me to sell their "moss" balls.

Chondrus crispus (Irish moss or carrageen moss) is actually a species of red algae that flourishes on rocky boulders along the Atlantic coast of North America and Europe.

Lichopods

Lichopods (family Lycopodiaceae) are vascular plants sometimes referred to as fern allies. Like ferns, they have the distinction of reproducing through spore dispersal rather than flowers. They have true roots and scalelike leaves that

Moss confusion among hobbyists

Beyond the world of gardening, confusion from name generalizations is common among aquarium, terrarium, and vivarium enthusiasts. Some plants called mosses used in aquascapes are indeed aquatic mosses such as *Taxiphyllum barbieri* (Java moss), *Vesicularia dubyana* (Singapore moss, triangular moss, mini moss, willow moss), and *Vesicularia montagnei* (Christmas moss). But some "moss" plants featured in aquaria are algae, not mosses. Often suppliers of aquarium or vivarium mosses do not distinguish between true mosses, lichens, lichopods, and algae.

To complicate the matter further, the majority of aquarium suppliers are misinformed or unaware of scientific names for their products, and multiple moss species are called by the same common name. For instance, *Leucobryum*, *Atrichum, Plagiomnium*, and *Dicranum* species are all called frog moss. Frankly, frogs, salamanders, lizards, and turtles would be happy with a wide range of moss species. It is more important for amphibian and reptile hobbyists to choose an appropriate moss that suits the moist environment than one that is arbitrarily called frog moss.

are spirally arranged in a pattern similar to how moss leaves wrap around their stem. Many common names—including club moss, shining club moss, shining fir moss, turkey brush, running ground pine, running ground cedar—are associated with a variety of lichopod species.

Selaginella plant species resemble certain mosses but are classified as lichopods. *Club moss* is a term frequently used for all species in this family of plants. Other common names include spike moss, blue spike moss, peacock moss, and golden moss. *Selaginella* is another popular nursery plant sold in the United States. I have planted a *Selaginella* species native to North Carolina in my fairy garden. It maintains its neon green through normal growing seasons but usually dies back to brown or black during the winter months. Since I find this stage quite ugly in comparison to cold-tolerant mosses, I pull off dead branches and leaves of *Selaginella*. Like other perennials that die back, it rewards me each spring with expansive new growth, once again creeping along rocks and tree stumps.

Selaginella plants that have contrasting foliage in golden hues, blue overtones, and red hints are *Selaginella kraussiana* 'Aurea' (golden moss), *S. uncinata* (peacock moss, peacock spikemoss, blue spikemoss), and *S. erythropus* (ruby-red spikemoss). Like bryophytes, spikemosses grow all around the world, including Braun's spikemoss in China and Krauss' spikemoss in Africa and the Azores.

Besides *Selaginella*, many other club mosses exist in the family Lycopodiaceae, some of which keep their deep green color all year long. In my neck of the woods, *Huperzia lucidula* (shining club moss) is evergreen with fuzzy fingerlike plants that sneak from under rocks or decaying logs. At sites where logging has occurred and new-growth pine trees are present, the entire forest floor is covered with what we call running ground cedar or turkey brush. *Diphasiastrum digitatum* is the scientific name for this trailing lichopod, which has a beautiful circular leaf branch arrangement with impressive spore spikes rising above. If you want to grow these particular lichopods in a landscape, it can be a challenge. The fibrous roots along the trailing runners are insufficient for establishment in new locations. Hint: It is necessary to use the tip end of the extended runners if you want to succeed with growing this lichopod species.

Lichopods are highly valued for their medicinal properties, particularly in Asian cultures. The evergreen properties of certain club mosses make them prized for decorative purposes in the floriculture industry. Once again, harvesting practices to meet public demands are having a negative impact on native populations.

TOP Club mosses are lichopods, not bryophytes.

Some club mosses, such as this *Huperzia* species, resemble bryophytes (true mosses).

mosses for gardeners

25 BRYOPHYTES TO KNOW AND GROW

Wondering which mosses to plant? This section introduces gardeners to a selection of recommended species. Quite a few mosses are native to all parts of the United States and other places in the world. Other species, closely related in the same genus, may vary in geographic distribution.

If you are willing to adapt your environment with soil pH adjustment and/or supplemental watering, you have many options to consider as horticultural choices for your moss garden, lawn, patio, or water feature. Keep in mind that the mosses included here are just a sampling of species with proven success in manipulated landscapes and gardens.

Identifying mosses can be a challenge. Bryologists use microscopic details such as arrangement of cells, leaf margins (edges or borders), and/or shape of the spore capsule to identify bryophytes. Since the majority of moss gardeners are not inclined to use a microscope, we will investigate characteristics you can see with the naked eye and the use of a handheld close-up lens (jeweler's loupe). With time, you can develop a keen eye for botanical characteristics, growing habits, shape, and color. Use your sense of touch to learn the nuances regarding density of colonies or loft of mats. If you want to delve deeper into identification using specific characteristics, keys, and microscopes, you should refer to field guides and ID references available in print or on the Internet.

A note about soil types

Many mosses prefer acidic soil, but some do not, so I give the preferred soil type for mosses listed here. I will say more about testing your soil pH in the chapter on planting mosses, but for now suffice it to say that the lower the pH number of your soil, the higher the acidity.

» Acidic soil has a pH of less than 6.

» Calcareous soil, which contains calcium carbonate, has a pH greater than 7 and includes soils dominated by limestone or dolomite rock.

» Alkaline soil has a pH greater than 8.5 and includes clay soils and those infused with sodium carbonate.

PAGE 100 *Climacium americanum* is my favorite moss. Beyond year-round appeal, this species is an excellent grower from plant fragments.

Anomodon rostratus

common tree apron moss, yellow yarn moss

plants ½–¾ in. tall or long; leaves 1⁄32–⅛ in. long

calcareous or alkaline soil, sometimes acidic

Anomodon rostratus is a pleurocarpous moss, meaning it grows sideways. This species likes alkaline soils, in contrast with acid-loving mosses. It is widespread throughout the world, frequently occupying a place on bark at the base of trees (especially oak, hickory, or pine) or living on cliffs and boulders rather than on soil. In Asia, another species in the same genus is more prevalent, *Anomodon longifolius*.

Plants can be dark green to a brighter green with yellow overtones. When dry, *Anomodon rostratus* changes only slightly in color and gets a dull appearance. Wisconsin moss gardener Dale Sievert says, "While *Anomodon* does get a bit dull, it still is not bad looking." He favors using this moss, which grows in abundance nearby, most often on limestone rocks and rarely on granite. The spore capsules are covered by a beaked operculum, a fact referred to by the species name *rostratus*, which is Latin for "beaked, curved, or hooked."

Atrichum angustatum

star moss, little star moss, wavy starburst moss
plants ¾ in. tall or long; leaves ⅛–¼ in. long
acidic soil

Atrichum angustatum and *Atrichum undulatum* are in the same genus but differ in size and growing conditions, particularly their preferred soil quality and sun exposure. Both of these recommended species have a relatively consistent plant height and present a smooth horizontal expanse of greenery, ideal for emulating the effect of a lawn; colonies of these mosses form mounds only when covering rocks or other raised soil humps. Both have star-shaped tops, but *A. angustatum* is the smaller species, grows lower to the ground, and tolerates the worst of nutrient-poor soils. It is frequently found growing on roadsides as well as on the exposed roots and soil of overturned trees. Colonies growing in sandy conditions tend to fall apart during retrieval or planting, so be careful while handling. Keep any loose gametophytes because *Atrichum* grows well from fragments.

The main leaf color is a medium green, but the male cups are a showy orange, and this display is enhanced because males group together in separate nearby colonies (the plants are dioicous). *Atrichum angustatum* grows in shade or sun and makes a good choice in the cracks between patio pavers or stones. Despite visual variances among *Atrichum* plants, sporophytes look remarkably alike to the naked eye. Both *A. angustatum* and *A. undulatum* offer a bronzy brilliance during impressive sporophytic stages. The slightly curved, elongated, cylindrical mature capsules are dark brown with a contrasting white covering (epiphragm) over the capsule opening.

Atrichum undulatum

big star moss, crane moss, crown moss, starburst moss

plants ½–2 in. tall or long; leaves ¼–½ in. long

acidic soil

This moss has star-shaped gametophytes several times the size of *Atrichum angustatum* gametophytes. *Atrichum undulatum* offers exceptional beauty, with translucent green leaves a few shades lighter than those of *A. angustatum*. It is normal for the plants to go through a rusty brown stage. Even though it appears that disaster has occurred, new growth of almost neon green will eventually emerge. Frequent and prolific sporophytes are common. Before full maturity, the golden brown capsule has a long beaklike tip. The plants can be dioicous (separate male and female colonies) or monoicous (male and female reproductive organs living in the same colony or on the same plant).

While *Atrichum undulatum* tolerates partial sun, it thrives best in shady locations. It prefers wetter, humic soils in hardwood forests rather than the less desirable soils preferred by *A. angustatum*. This species is strongly recommended for a moss lawn where a high tolerance of foot traffic is needed. Like its relative, this *Atrichum* works well for stone paths or patios, too. Both of these upright growers are exceptional candidates for fragmentation planting methods. *Atrichum undulatum* can be aggressive, with rapid growth achieving full coverage in six months to a year. I have seen *A. undulatum* outperform other fast-growing mosses such as *Thuidium* and *Hypnum*, eventually crowding them out. Gentle finger raking of well-established colonies yields a wealth of small fragments. A vigorous hand swooshing freshens up browned-out older growth, encouraging and stimulating new growth.

Aulacomnium palustre

ribbed bog moss

plants 1¼–4 in. tall or long; leaves 1⁄16–3⁄16 in. long

acidic or alkaline soil

When it comes to a spongy-feeling moss, *Aulacomnium palustre* takes the prize. It can grow in mats 2–8 inches thick that hold together by subtle interconnectivity. In nature, it covers roadside cliffs that offer a consistent trickle of water from springs uphill. It also grows in swampy habitats such as bogs, marshes, and fens. *Aulacomnium* tolerates a range of soil pH levels and is considered cosmopolitan because it can be found growing around the world.

To describe the glowing green appearance of *Aulacomnium palustre* as neon green or chartreuse is almost an understatement. Closer inspection shows the leaf has a distinct main rib (costa) with a reddish brown tint. When this species dries out, it becomes somewhat lackluster because of microscopic bumps on cells in the leaves. Sporophytes are reportedly rare on this species, but I've seen them. Asexual reproduction occurs with gemmae—leafy stalks rising above the colony with globelike vegetative propagules.

The thick, bulbous mats can be used to imitate mountain ranges or rolling hills in moss landscapes. Preferring extreme wetness, this moss species likes the splashes of waterfalls, rain gardens, and bogs. While sitting or walking on lofty *Aulacomnium* colonies is quite comfortable, the plants may show signs of stress after too much squashing. However, as long as new green tips are present or reappear, the plants should rebound with vigor. One thing is for sure—if you sit on these moss cushions, you will walk away with a wet behind!

Bartramia pomiformis

apple moss

plants ¾–2½ in. tall or long; leaves 3⁄16–¼ in. long (edges curl inward)

acidic or calcareous soil

Bartramia pomiformis is a cushiony, fluffy moss species in a light medium shade of green, sometimes yellowish. It has curving leaves that are somewhat dull because papillose (having small bumps). Short sporophytes (less than an inch tall) are distinctive, with round capsules that resemble little green apples in their early stage (*pomiformis* means "apple-like" in Latin) and turn reddish at maturity. Plants are loosely connected in mound shapes, usually 2–5 inches tall. *Bartramia* prefers rocky ledges or sloping hillsides that offer plenty of moisture and shade, and can be found growing in such locations around the world.

This species was named by Johann Hedwig after Pennsylvania naturalist and nurseryman John Bartram, who collected all types of plants in the eighteenth century and was in direct communication with European botanists Linnaeus and Dillenius. This acid-soil-loving moss species prefers moist conditions; plant it around water features. It tolerates partial sun but really prefers shady spots. At my house, *Bartramia* goes into its sporophytic stage in the winter months. When the snow melts, I am always delighted to see its apple-shaped capsules.

Brachythecium rutabulum

rough foxtail moss, cedar moss

plants ¾–2½ in. tall or long; leaves 3⁄16–¼ in. long

acidic or calcareous soil

Brachythecium rutabulum grows in horizontal mats that are somewhat shaggy and always glossy in appearance. It has both creeping and arching upright stems with ovate-triangular, smooth-edged leaves in the green-to-yellow range. Sporophytes are rust colored, with the distinction of having teeny bumps running the entire length of the ½-to-1-inch seta (spore stalk). The brownish capsules have been described as macaroni shaped. This plant is widely distributed around the world although it can be a challenge to identify with its small capsules and the fact that plants are often sterile (without sporophytes).

During the photosynthesis process, *Brachythecium* can deposit calcium carbonate, and thus it can be considered a contributor to the formation of rocks. It tolerates a range of exposures—shade or sun—in natural habitats and varies little in appearance whether it is wet or dry. A moss for all seasons, *Brachythecium* is a good choice for challenging locations in your garden. Plant it where all other plants have struggled to survive and use it to enhance patio pavers or stone paths.

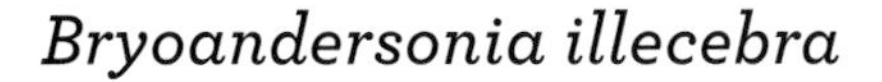

Bryoandersonia illecebra

spoon moss, cup moss, worm moss

plants ¾–2½ in. tall or long; leaves 3/16–¼ in. long

acidic or alkaline soil

The loosely connected matlike colonies of *Bryoandersonia illecebra*, an acrocarpous (upright-growing) moss, can have bright green or yellow hues, depending on sun exposure—the more sun, the more golden the gilding. Tiny leaves cup around the 1-to-2-inch plant stems. Some people think the tubular gametophytes resemble a mass of glimmering yellow-green worms. This species can live in deep shade along a stream bank in bands that seem to beam in the filtered sunlight. In brighter spots, thick mats adorn rocks like sparkling golden crowns. *Bryoandersonia* can grow on rock substrates, tree bases (particularly oak or hickory), or directly on either acidic or alkaline soils. This moss tolerates extremely hot seasons and dry conditions.

Of special note, *Bryoandersonia illecebra* is endemic to North America, where it grows in several regions, and does not grow elsewhere. It was named by Harold Robinson after the late Lewis Anderson, co-author of *Mosses of Eastern North America* and distinguished Duke University botany professor. Sporophytes emerge from the colony for an ombré effect or gradation of colors as they mature. This species makes a decorative garden accent. Be wary of planting it in an area with heavy foot traffic because colonies dislodge easily. Also, you may need to repair bird or critter damage.

Bryum argenteum

silver moss, sidewalk moss, crack moss, asphalt moss
plants ⅛–½ in. tall or long; leaves 1⁄32 in. long
acidic, calcareous, or alkaline soil

Bryum argenteum is a velvety moss with gem-like qualities. The deep green color and silver patina create intrigue, with further pleasure derived from its incredible softness to the touch. Considered a weedy moss, *B. argenteum* grows everywhere—literally around the world, in urban and natural settings alike. Labeled cosmopolitan and ubiquitous by scientists, it grows on sand, cinders, gravelly areas, railroad tracks, and all types of pavements, including bricks, asphalt, and concrete. Often it finds its way to live on roofs. You have likely encountered this sun-tolerant moss many times while hiking sunny trails or walking down the street. The silvery leaf tips of this hardy species are its most distinctive and recognizable feature when plants are sterile (no sporophytes).

The glittering silver appearance is due to the fact that *Bryum argenteum* lacks chlorophyll in cells near the tips of the leaves (apical cells), making the leaves look whitish. Smaller leaves, tapering at the tips, emphasize the effect. When the moss is dry or dormant, the white can become dull and gray. To identify the differences in plant shape between *Bryum argenteum* and *Ceratodon purpureus* (a companion species), you will need to use a close-up lens. Short sporophytes are common in my neck of the woods, but in the Northwest this moss rarely gets sporophytes. Vegetative reproduction takes place via fragmentation. The white leaf tips break off to reproduce asexually.

Even though *Bryum argenteum* can thrive in bright or partial sun, it requires the moisture found in areas lacking good drainage or that receive rainwater runoff from buildings or nonpermeable surfaces. At the same time, it needs to dry out periodically. A consistent supplemental watering regime is not necessary for this species, just an occasional drink. In fact, I've had *Bryum* colonies shrink into nothingness from oversaturation or too much shade. Considered especially resistant to the harmful effects of pollutants, *B. argenteum* continues to thrive when drenched in the by-products of combustion engines. Plant this moss in places where the heat index is high. It makes a good choice for outdoor living spaces with pavers or stones. Tucking *B. argenteum* into the nooks and crannies of walls adds instantaneous antiquity.

If you want to try planting using fragmentation techniques, crumble up colonies with soil still attached; rub colonies vigorously in your hands and tear them apart; pulverize in a blender only if you wish; then make up a mud pie mixture. Apply to rock substrates or on soil to fill in cracks between stepping-stones.

Ceratodon purpureus

fire moss, purple-horned toothed moss, purple moss

plants ⅛–½ in. tall or long; leaves 1⁄16 in. long

acidic soil

Spectacular crimson sporophytic displays make this tiny species a flamboyant moss. Yet, more often moss gardeners want it for its deep to bright green appearance and the velveteen texture of the colonies. *Ceratodon purpureus*, with its royal attributes, is a universal species. It grows in all types of climates and habitats, all around the world—from the frozen ice of Antarctica to the blistering sidewalks of New York City. Though it is typical of acidic soils, it tolerates a range of soil pH. As upright growers, the super small plants barely rise above the ground. Rhizoids grow into soil to about ½ inch deep.

Ceratodon seems to prefer harsh conditions and substandard sites. In fact, this moss is a sun worshipper. Although it tolerates cold temperatures, it is partial to hot sites such as the cracks of concrete sidewalks and corners of asphalt parking lots where sweltering heat radiates from the surface. It doesn't mind baking in the sun, even living on roasting roofs and the sizzling surfaces of boulders. Niches offering periodic moisture are favored over consistently moist conditions; do not overwater this species or you will be disappointed. *Ceratodon* is ideal for stone patios and paths, rock walls, and green roof applications. For fragmentation planting, follow the procedures recommended for *Bryum argenteum*.

The genus name *Ceratodon* comes from a Greek term referencing the forked peristome teeth (around the capsule opening) that resemble the horns of an animal; *purpureus* is Latin for "purple." Apparently in classical terms, purple is a reddish color. Often in the United States, we think of purple in more bluish terms. I consider the setae of the sporophytes quite red in color, not purple. After the cerise-colored sporophytes fade away, their mother plants turn dull brick red for a short period. As the life cycle of the colony continues, these dreary-looking sections again transform with the emergence of intensely green new growth.

Climacium americanum

tree moss

plants 1½–3¼ in. tall or long; leaves ¹⁄₁₆–⅛ in. long
acidic or calcareous soil

Small is a relative term when it comes to the miniature world of mosses. As mosses go, *Climacium americanum* is a giant genie among tiny tots. Its upright growth towers above other species. The upper plant growth is most impressive as it matures from small, spherical-shaped, emerald Christmas trees into larger spreading branches as wide as a few inches. These lush topknots display a range of colors from the brilliant greens of younger plants to the duller olive green and brown stages of older growth. Delightfully, these transitions are simultaneously occurring within colonies all the time.

In fall, the leaves on the tips of plants acquire an incandescent orange sheen as if a fairy had dabbled a bit of paint to add her sparkling touch. Sunshine yellow replaces the green tones when this species is located in sun-drenched exposures that receive adequate rainfall and/or supplemental watering. As an exception to the general rule about poor soil conditions, it thrives in rich loam similar to a forest floor. Moisture is a key factor leading to consistently healthy plants.

Although *Climacium* thrives in shady locations with high moisture retention, even soggy or swampy spots, it can live in the sun as well. It invades grass lawns, a phenomenon I have noticed in North Carolina. In the Northwest, the other *Climacium* species prominent in America, *Climacium dendroides*, is considered a weed in grass lawns. Excellent at vegetative reproduction, *Climacium* rarely has extensive sporophytic displays. But when it does happen, several sporophytes (with cylindrical capsules) originate from the same tip like a tuft of feathers. In contrast, most acrocarpous mosses have only one sporophyte per plant; *Rhodobryum* is similar to *Climacium*, with headdresses of multiple sporophytes.

With a combination of upright growth and tough horizontal rhizomes, individual plants link together in long lines. This linear growth feature is not apparent when colonies mass together and intertwine into groups. This is the privileged view of a moss gardener. Keep colonies intact for focused impact or gently coax them apart into strings with "little trees" growing a few inches apart. Either way, you can create a forest feature in miniature with massive appeal.

Plant *Climacium* moss colonies for erosion control, and remediate waterlogged areas with these thirsty mosses, which can handle rushing water from periodic rainfall, flash floods, or urban runoff. *Climacium* colonies live on boulders and soil substrates in mountain fens and bogs where similar conditions exist. When planting colonies, supplement compacted, clay-based soils with decomposed leaf and pine needle litter. Furrow up the ground a bit with a rake, adding light layers of supplemental soil

mix underneath and over the rhizoids. Make sure to reposition all "trees" in an upright position when finished.

This species is stupendous when it comes to growing well using fragmentation planting methods. All plant parts can regenerate; even seemingly dead areas can sprout new green from along the rhizoids, plant stems, branches, and leaf tips. Given the tensile strength of the stems and rhizoids, scissors are necessary to cut plants into ½-to-1-inch pieces as source fragments. Attachment to soil happens within several weeks or months, and visible expansion can be impressive within six months to a year. Thick distribution of fragments, walking on the moss fragments, and supplemental watering will yield the best results.

Dicranum scoparium

mood moss, windswept moss, broom moss, footstool moss, cushion moss

plants ¾–3¼ in. tall or long (but I've found examples as tall as 8 in. with multiple heads); leaves 3⁄16–⅓ in. long

acidic soil

With widespread distribution across North America and around the world, *Dicranum scoparium* and other *Dicranum* species are popular choices for moss gardening. Slightly curved leaf tips give *Dicranum* a windswept appearance. Some people compare this curvature to the worn, tattered ends of broom straw. *Dicranum* colonies somehow bond together without elaborate intertwining. The cushiony mounds of this upright (acrocarpous) moss may be somewhat flat or appear as huge rounded hummocks, depending on the age of colonies. It is a super soft moss and makes great footstools at benches. *Dicranum* mosses can often be found growing under pine trees on hillsides and can live on soil or humus, sand, rock, the bases of trees, and rotten wood.

The color is normally a rich hue of green in the shade and a darker, denser green in partial sun. During seasonal changes in leaf canopy and more intense sun exposures, it can shift in color to gain a yellowish gleam. When harvested during its best color stage, it retains that color for months, even when dried out. Unfortunately, this feature makes this moss popular in the floriculture industry and for craft projects. On the other hand, this characteristic is attractive to moss gardeners who want a moss species that won't go through brown stages (sometimes considered ugly) during its normal life cycle or after sporophytes have died back. Golden sporophytes with almond-shaped capsules light up green colonies like glowing candles.

Most of the time, this moss species grows better in shade, but it can survive in some sunny situations. While preferring acidic conditions, *Dicranum scoparium* tolerates a broad range of soil pH. It is quite sensitive to prolonged wet feet—which can cause colonies to suffer significantly and even die—and often does better on hillsides where better drainage is possible. A yellow spot in the center of a colony is a sign of ill health. Plant *Dicranum* under trees that shed needles or add a layer of pine needles underneath colonies. In my experience, this can be a picky moss. Too much moisture causes it to sog out and too much sun can cause it to turn a sickly yellow and dry out, with negative consequences. *Dicranum* seems to be susceptible to frequent visits by birds and critters that toss the colonies around, necessitating repairs.

Entodon seductrix

seductive entodon moss, toothpick moss, cord glade moss

plants ¾–1½ in. tall or long; leaves 1⁄32–1⁄16 in. long
acidic, calcareous, or alkaline soil

With the ability to attach quickly to just about any substrate imaginable, *Entodon seductrix* offers beauty for shade or sun exposures. A particularly appealing characteristic is that *Entodon* gets superbly shiny. As if coated in a metallic paint, it shimmers with an iridescent quality, whether green, golden, or brownish. This glossy moss shifts color depending on the amount of sun it receives. Particularly pointy leaves overlap each other. Bronze capsules emit yellow billowing clouds of spores, an occurrence that may happen more than once a year. The genus name, *Entodon*, refers to the way the peristome teeth are embedded within each capsule like teeth in your mouth. My nickname for this species is shiny, sexy moss.

Classified as a pleurocarpous moss, *Entodon* grows sideways with a flatter appearance than most mosses. If you don't want any height, this is the moss for you. And you cannot go wrong in selecting a suitable location—this moss can grow just about anywhere and can thrive in excessive heat. It grows well on clay-based soils in open woodlands. The rhizoids hold tightly to rough surfaces like concrete or asphalt roof shingles, less so to soil. Accepting shade or sun, it makes an excellent choice for moss gardens, living walls, and green roofs.

Although plants interconnect, the colonies fall apart if not handled with care when planting. Spread fragments over pebble paths to achieve a hint of green. Since you can easily separate these small plants from each other by hand, there is no need for any elaborate preparation when using fragmentation planting. Do not even bother with making a mud pie if planting on soil. Distribute *Entodon* fragments over the surface and then water and walk on them. Despite the speed at which it will attach, do not expect fast expansion. Colonies seem to gain more loft before they start to spread out very far.

Fissidens dubius

plume moss, flat fork moss, pocket moss, maidenhair pocket moss

plants ¾–2 in. tall or long; leaves 1⁄16–3⁄16 in. long

acidic or calcareous soil

Fissidens dubius wears its own special crown of distinction in the world of mosses. The rich green leaves are all on the same plane, giving a flat appearance. In double rows, they run up the stem resembling tiny feathers. However, what really sets this species apart from its cohorts is that each new leaf fits into a pocket on the base of the previous leaf. No other genus has this unique botanical feature. The building block leaves also serve as reservoirs for water accumulation. Large and small teeth are located along the leaf margins. Use a loupe to see all these botanical idiosyncrasies.

Naturally growing along streams and waterfalls, this attractive moss thrives in areas of high humidity and frequent rainfall. Take a hike into the deep woods and you will see it growing on logs, stumps, rocks, and forest soil. Primarily a shade-loving species, it will grow in partially sunny spots, too. Filtered shade allows this moss to attain a deeper green tone. Sporophytes arise from the middle of the plant with a blaze of red—seta and capsules alike. The spores are encased in cylindrical tubes similar to the shape of *Atrichum* capsules. More than seven hundred *Fissidens* species can be found worldwide; other *Fissidens* species native to North America include *F. andianthoides* and the aquatic species *F. fontanus*.

Fissidens is an excellent starter moss, easily growing from fragments. Rhizoids connect into the soil effectively, and even this small moss helps with erosion issues. Place colonies in a contiguous manner to keep soil from washing away and to slow down stormwater. Use your sense of design in deciding where to plant this moss species. In a miniature landscape, *Fissidens* adds a certain panache with its textural appeal.

Funaria hygrometrica

twisted-cord moss, fire moss, Cinderella moss, predictor moss, charcoal peddler

plants ⅜–1¼ in. tall or long; leaves 1⁄16–3⁄16 in. long

alkaline soil

Fanciful and bountiful sporophytes are the hallmark of this pioneer moss. From the charred wood left in the wake of destructive forest fires, *Funaria hygrometrica* springs forth within less than a year to begin the regreening of the barren land. A short moss, it may go unnoticed until the parade of sporophytes begins. Even from a distance, spiky green resembling new grass yields to scarlet bands blending into golden waves that are eye catching for several months. The gradation of colors as the spores mature inside pear-shaped capsules is an impressive spectacle. In their final stages, the sporophytes darken to a deep crimson-copper color.

But this moss's glorified state is short lived. Its setae (sporophyte stems) begin to contort and twist together in old age—hence the common name twisted cord moss and the genus name *Funaria* derived from the Latin *funis*, meaning "rope." Within two to three years, this transient moss will have faded away, overshadowed by the growth of other vascular plants. Further, it is self-inhibitory, avoiding growing in the same place again. So although you could plant it where fireplace ashes have been discarded, it will not last or provide the longevity of other species. The fleeting beauty is not sustainable, no matter how much you try.

Recently I noticed this moss growing in the gravel parking lot of a hilltop nursery. Not just found at fire sites, *Funaria* is common on disturbed soils around the world. Yet it must be tolerant of huge doses of fertilizer since it lives on the soil of potted plants in greenhouses.

Hedwigia ciliata

Medusa moss, white-tipped moss

plants 3/8–5 in. tall or long; leaves 1/16–3/16 in. long

acidic soil

For the younger set, Hedwig is the well-recognized name of the owl in the Harry Potter books. But this bryophyte found around the world is actually named after the father of bryology, Johann Hedwig. With cascading green tendrils, *Hedwigia ciliata* looks extraordinary when in its wet state. Quite the opposite is true when it is dry—it gets dull, grayish, and stringy looking. But upon rehydration, it bounces back quickly, a candidate for the misting moss magic trick.

The plants group together in a rhizoidal foot, slightly fanning out with lighter green–whitish ends curling up like a hairdo. On flat surfaces, the moss tends to grow in cushionlike circular colonies with tendrils radiating out from the center. Some moss gardeners consider this a lovely addition to rock gardens with the frosty appearance of the lighter, whiter tips. Heat tolerance is another characteristic of this species I call dreadlock moss. It frequents boulders in shady or sunny natural settings, while its city cousins occupy walls, bridges, and rooftops. I never fail to be amazed that this lightweight moss doesn't blow off my roof during high winds and heavy thunderstorms.

The sporophytes of *Hedwigia* are inconspicuous and rarely noticed. However, as a curious moss gardener with a loupe hanging around your neck, you can peer into the depths of the delicate branches to find them. The sporophytes have really short stalks that are attached to specialized leaves deep within the colony.

Hedwigia ciliata foot

OPPOSITE *Hedwigia ciliata* colony

Look closely and frequently so as not to miss this reproductive occurrence.

Periodically, entire colonies of this moss go through a stage that is cantaloupe color. However, I have no explanation for this transition. It does not seem to be triggered by sun changes that affect the colors of other mosses. The first time I saw this orangey stage, I was worried that the colonies were sick. But since the orange always changes back to green, I have accepted this transition with wonderment. Maybe someday a bryologist or horticulturist will research why it occurs.

When planting *Hedwigia* colonies, keep them oriented in the same direction and layer them as if placing shingles on a roof. This is critical to get the foot to attach. If you want to decorate walls or boulders, add a bit of soil mud pie mixture to the crevice where you want *Hedwigia* to grow.

Hylocomium splendens

feather moss, stair-step moss

plants 1–2 in. tall or long; leaves 1⁄16–1⁄8 in. long

acidic soil

A larger, somewhat coarse moss, *Hylocomium splendens* has a feathery appearance. The fernlike plants stair-step upward along upright stems that can be stiff and wiry. In the Northwest, *H. splendens* doesn't have stair steps but irregularly shaped branches. Colonies interweave in a carefree fashion into thick blankets. Plants are medium green and assume a dull, almost faded green color as colonies dry out. Considered an occasional occurrence, sporophytes provide a bronzy brown glimmer above the surface of the luscious moss mats.

This widespread species grows in montane habitats on soil, humus, and logs. It is an extremely fluffy moss that makes a soft bed or interesting focal contrast to other mosses. *Hylocomium* grows best in shade. As with *Climacium*, rough up the soil surface and add a thin layer of supplemental soil mix.

Hypnum curvifolium

fern moss, log moss, sheet moss, flat-tufted moss, brocade moss, carpet moss, curly-leaf fern moss

plants ¾–1½ in. tall or long; leaves 1⁄32–1⁄16 in. long

acidic soil

The genus *Hypnum* includes more than eighty species, several of which are indigenous to North America. *Hypnum curvifolium* looks like dwarf ferns woven or braided into fairy carpets. A sideways grower, *H. curvifolium* thrives in moss gardens. Single plants curl toward the ends and knit together into mats, creating colonies that can extend across large expanses as their branching gametophytes reach for new ground. Branching within the cluster increases the overall loft, resulting in thicker, spongier walking delights. Green for a majority of the year, this species will go into golden grandeur in response to leaf canopy changes. The lighter-colored leaf tips stand out distinctly against the monochromatic background. It is prolific in terms of sporophytic reproduction, and displays during this stage can be divine.

The name *Hypnum* is derived from a Greek word meaning "sleep," which was applied in antiquity to mosses and lichens used for medicinal purposes. Like other *Hypnum* species, *H. curvifolium* prefers acidic soil but can live on a variety of soils, rocks, and trees. It is more resilient and flexible in terms of climatic conditions than its common companion species *H. imponens*. Consider planting *H. curvifolium* in your moss garden, via fragments or whole colonies, for its tantalizing textures.

Hypnum imponens

fern moss, log moss, sheet moss, flat-tufted moss, brocade moss, carpet moss

plants ¾–1½ in. tall or long; leaves ⅓₂–⅟₁₆ in. long

acidic soil

Hypnum imponens is a universal species found growing extensively on decaying logs and in woodland habitats throughout North America and the world. The fernlike shape of the individual plants melds together in colonies for a brocade appearance. A horizontal grower, *H. imponens* forms mats that can be thin or several inches thick depending on environmental conditions or colony age.

This medium-green moss can sometimes get a darker tint in dense shade or attain a bright golden, even lemony yellow color from increased sun intensity during early spring or late autumn (right before new leaves emerge or after they fall off). Like *Dicranum*, if harvested when brilliant green, it will retain this color when dry for long periods of time, making it a desirable species for crafts and floral arrangements. The stems have an orangey brown tone visible without a close-up lens. Using your loupe, you can see the colorful stem hidden under leaves that curve in the same direction. Also, observe that this species does not have a midrib (costa), a characteristic common to most mosses.

Hypnum handles dry conditions better than *Thuidium*, the other popular fern moss. Its appearance changes very little. *Hypnum* often prefers good drainage and seldom likes consistently soggy conditions. In fact, it turns brown and sickly looking when it stays wet for extended periods, which I refer to as sogging out. Use *Hypnum* in all types of shade garden applications—focal features, waterfall adornment, or as decoration for rock walls. This genus dominates some of the most impressive moss lawns in this country.

Red flag warning: Sometimes *H. imponens* can be a persnickety moss to grow. For success, plant directly on soil with good drainage and a pH of 5.5. Interleaf the edges to help colonies connect to each other. Use twigs to pin edges together. If you are willing to wait for *Hypnum* to grow in, tear colonies apart into small pieces and spread the fragments on the ground.

Leucobryum glaucum

pincushion moss, cushion moss, white moss

plants ¾–3½ in. tall or long; leaves 1½–3¼ in. long

acidic or alkaline soil

This cushion-shaped moss species and its close relative *Leucobryum albidum* can be a lighter, more subtle shade of green than many bryophyte species. The whitish tone (*leuko* means "white" in ancient Greek) is due to a botanical departure from the typical one-cell-layer leaves of other species. In this species, hyaline and chlorphyllous cells are interwoven. As top layers dry out, they lose their color. Upon rehydration, *Leucobryum* soaks up water like a sponge. Colonies retain moisture quite effectively, so it takes this species a long time to dry out. Competing moss species grow on these moist mosses as a preferred substrate. Spores cast to the wind from various moss genera land in the hospitable and nurturing environment of *Leucobryum* mounds.

Leucobryum species grow in all types of habitats and are considered widespread in terms of geographic distribution. Most often, they grow on shady forest floors, on logs and stumps, and at the bases of trees and old-growth shrubs (such as rhododendron). It has been cited that *L. albidum* is one of the few kinds of mosses that grow on pine bark; further, I've seen both *Leucobryum* species growing at the beach in the sand—in the sun.

Leucobryum glaucum plants are larger than *L. albidum* plants. When separated from the mother colony, the latter is not much bigger than a stud earring, whereas the former can range up to finger-sized plants. *Leucobryum glaucum* colonies do not grow as tightly or densely as *L. albidum* colonies. Sporophytes, short in stature, occur in significant numbers on *L. albidum* and infrequently on *L. glaucum*. Especially effective at vegetative reproduction, both *Leucobryum* species have leaf tips that break off, creating white coverings on the mounds. Some blow away to new locations and form perfectly round pincushions as they begin to grow. Others are absorbed into the inner sanctum of the colony, thereby "reseeding" and increasing the colony size via new growth.

When you plant mounds of *Leucobryum*, provide decent drainage by adding an underlayer of pine needles. Both *Leucobryum* species can live in shade or sun. If they are planted in the sun, you may want to provide periodic drinks more often. It is easy to tell when *Leucobryum* mounds are thirsty because the leaves get whiter. A quick quench and they green up before you have finished watering the rest of the garden. Overwatering *Leucobryum* results in soggy enclaves that transition into host locations for other aggressive moss species. Crumble colonies or cut with scissors to produce fragments for planting.

TOP *Leucobryum glaucum* colony

BOTTOM *Leucobryum albidum*

Mnium hornum

lipstick thyme moss

plants ¾–2 in. tall or long (males a bit taller);
leaves ⅛–⅕ in. long
acidic soil

Mnium hornum is a shade-loving moss species that grows in wetter areas around the world. Its small oval leaves cascade intricately along a single plant stem that doesn't branch. Its light shade of green signals its see-through leaves. When viewed through a loupe, the forked edges of the leaves have a double-tooth aspect. Hold the plant specimen up to the light and you will be able to see two tiny pinpricks. In contrast, the leaves of *Plagiomnium* species have a series of single points instead of two points like *Mnium* (an ancient Greek word meaning "moss").

Although *Mnium hornum* is considered an acrocarpous (upright-growing) moss, the female plants trail outward in a prostrate manner or drape down if growing on the vertical plane of a stream bank. Erect-growing male plants have slightly longer leaves and a rosette-style splash cup. (With male and female plants, this species is dioicous.) When dry, *Mnium* leaves shrivel up. The sporophytes are indeed splendid, with heads that nod as if in reverent prayer. Only a couple of inches tall, they seem to tower over the gametophytes. Spore capsules resemble the shape of a somewhat deflated football.

You might consider planting this moss in a feature section with other contrasting colors and textures. As a backdrop, it showcases other ferns and wildflowers you may include in your moss garden design. Like most *Mnium* species, it loves a good drink. Although water features are good candidate locations, this species grows well on patios or even on north-facing walls. It can tolerate partial sun, but you will realize best results in the shade.

Plagiomnium ciliare

many-fruited thread moss, saber-tooth moss

plants 1¼–2½ in. tall or long; leaves ⅕–⅓ in. long

acidic soil

Flowers on a moss? No way—that's a botanical impossibility. But when you examine a *Plagiomnium ciliare* colony, it appears to have green blossoms. In actuality, the delicate "petals" are leaves of the male plants, which are distinctively different from the females ("Vive la différence!"). In this case, the males form splash platforms at the top (apex) that indeed look like fairy florets. If you check out the males, you can see dark balls ready to splash out when a giant raindrop hits the cup. The female gametophytes resemble vascular plants. In contrast with *Atrichum angustatum*, with males located in separate colonies, males and females of *Plagiomnium ciliare* live together in the same colony ("Vive la joie de vivre!"). This close proximity aids in reproduction. As always, sporophytic displays offer special appeal.

The most alluring aspect of *Plagiomnium ciliare* (originally named *Mnium affine*) is the transparency of the leaves. The glistening green of the plants en masse becomes more translucent when you look closely at individual leaves. This species prefers rich, highly acidic soils to thrive. With natural habitats ranging from boreal forests to swamps and streams, this moss is a no-brainer choice for water features. Mainly a shade lover, *P. ciliare* can even be the showcased species in a moss lawn. It needs supplemental watering when used in moss gardens. When dry, the leaves curl inward and lose much of their lush appeal. You can broadcast fragments, cut in small pieces from live colonies, throughout target areas to gradually introduce this lovely moss.

Polytrichum commune

haircap moss, awned hairy cap moss, blue moss, blue hairy cap

plants 2–6 in. tall or long; leaves ¼–⅓ in. long
acidic, calcareous, or alkaline soil

I cannot sing enough praises of *Polytrichum*, a prolific moss genus found around the globe. *Polytrichum commune* is a warrior in the fight against erosion. Its strong and long rhizoids hold the soil as tenaciously as any vascular plant. Its ability to thrive when directly exposed to pollutants or toxins is extraordinary. It cares not whether it lives under the shady canopy of protective trees or on bare slopes with direct southern sun exposures. Further, this moss species grows in just about every soil substrate from red clay to white sand to black loam. Natural habitats include boreal forests, mountains, and prairies.

Polytrichum commune colony

Thick, opaque, pointy leaves are medium to deep green. New growth occurs on the ends of browning stems of older growth from previous years. You can see increments of this annual expansion. Plants may stand erect or drape in a reclining style. When drying out, leaves do not lose much color or shrivel up; instead, leaves tightly wrap around the plant stem in an effort to keep in moisture. This dramatic visual difference is easy to see. Be a moss magician and delight your friends by misting dry plants and demonstrating almost instantaneous rejuvenation. Your audience will be amazed to watch leaves unfurl in less than a minute.

This species' sporophytes are tall in comparison to those of other North American moss species. Setae (supporting stems) can be up to 4 inches long, topped with distinctive four-sided capsules when mature. *Polytrichum* gets its common name from the white, silky, filament-like hairs of the capsule covering (calyptra). Hence, the connection between the "hair" and the "cap," and the origin of the common name haircap moss. The eighteenth-century botanist Johann Dillenius called this species great Goldilocks. Although *Polytrichum* can grow from fragments, it is one of the most successful mosses to colonize new areas via spore dispersal. With up to one million spores per capsule, it often shows up in unexpected areas.

Individual plants of this acrocarpous (upright-growing) moss do not intertwine, but rhizoids bond the soil together to create colonies. *Polytrichum commune* is considered

the largest unbranched moss species in North America. These mosses are so hardy, they do not mind being walked or even stomped on and seem oblivious to damage from foot traffic. Sounds like *P. commune* has great potential for lawns and paths, doesn't it? When featured in lawns, this moss can be mowed to maintain a consistent height.

Plant versatile *Polytrichum* colonies anywhere you want additional texture and height. Male cups provide additional interest with their striking red, gold, and bronze colors. While taking advantage of its amazing visual appeal, you can benefit from its value in remediating erosion issues, too.

Polytrichum juniperinum has all the same qualities for moss gardening as *P. commune*. Visually, leaves offer a subtle shade of green with bluish hues tinged with reddish brown around the edges. This slight color distinction is due to the fact that the leaves slightly curl around the edges. Leaves are smaller, too, so the erect plant is skinnier than *P. commune*. Other relatives—including *Polytrichum ohioense*, *Polytrichum piliferum*, and *Polytrichum strictum* —differ in plant height and size as well as sporophyte height and spore capsule shape.

Polytrichum juniperinum

LEFT *Polytrichum commune* mounds

Rhodobryum ontariense

rose moss

plants ⅜–1½ in. tall or long; rosette leaves ⅟₃₂–⅟₁₆ in. long

acidic or calcareous soil

Rhodobryum ontariense resembles green flower blossoms in its hydrated state. The medium-green leaves form a rosette reminiscent of a tiny flower; otherwise, the stem is leafless. This is one of the species that dramatically changes from wet to dry. In the latter state, it may even go unnoticed because the leaves curl up into dark, dense green or brown nubbins. Upon rehydration, it unfurls before your very eyes, just like magic. This extreme sensitivity to moisture and the drying effects of sunshine dictate a shady location.

The leaves are almost see-through, like those of *Mnium* species. Sporophytes grow about 2 inches tall and emerge from the top crown of leaves on this moss in the same manner as *Climacium* sporophytes. The capsule turns a flame orange and has a dramatically drooping head. Plants grow upright along a strong sideways rhizome. From a distance, *Rhodobryum* colonies appear flat rather than mounded. Frequently, this species is misidentified in North America as *R. roseum*, which lives in Europe. Among the differences between the two species is that *R. roseum* has sixteen to twenty-one leaves in a single rosette, while *R. ontariense* has eighteen to fifty-two. The *Rhodobryum* moss in my yard is definitely *R. ontariense*—I've counted the rosette leaves!

Most often in my Appalachian mountain region, *Rhodobryum ontariense* grows at the base of trees or on top of rocks. Once I found it growing on a discarded piece of indoor-outdoor carpet in the woods! With growth habits similar to those of *Climacium*, *Rhodobryum* thrives in rich humus. In Europe, the farther north you go, the plants become sterile (without sporophytes). I'll admit it has been an exceptional occurrence for me to observe this species in its sporophytic state, but I have seen them.

Since the colonies are so incredibly beautiful and somewhat rare, I plant them intact. I cannot seem to make myself destroy any of them to experiment with fragmentation. I suspect it could work as long as you choose a densely shaded, sheltered, moist area. If you want royalty in your moss garden, *Rhodobryum* reigns as an opulent bryophyte choice.

Sphagnum palustre

peat moss, rose peat moss, rusty peat moss, ball white moss, drowned kittens

plants 3–10 in. tall or long; leaves 1⁄32–1⁄8 in. long

acidic, low-nitrogen soil

Sphagnum palustre is just one of more than two hundred *Sphagnum* species found around the world. Earlier I elaborated on the value of *Sphagnum* peatlands to our environment; now let's look at how to grow it in your moss garden. Individual plants do not really interlock in colonies but are grouped together in a loose, haphazard way. The uppermost branches create dense heads (capitula) on long tendrils. These flowerlike tops can be all sorts of colors. In the case of *Sphagnum palustre*, they are green in the shade or bronzy in the sun. Other species exhibit a range of greens, golds, browns, and reds; some are almost devoid of any color and look whitish. Colors are more intense in colder months and sunnier sites. *Sphagnum magellanicum* and *S. rubellum* are both quite spectacular with pale pink to deep rose tones. Some forty years ago, I slept under the twinkling stars on a thick bed of these spongy red mosses. Little did I realize how this experience would influence my mossy dreams in later years.

Sphagnum moss species definitely like wetter conditions and can be found growing in bogs or mountain fens. They have the ability to retain significant amounts of moisture—up to 33 percent more water than in their dry state—in specialized water-holding cells. After a heavy rain, *Sphagnum* colonies can be downright soggy. When plants are dry, the color fades but the cushiony characteristic remains. Line the edges of ponds, bogs, or stream banks with *Sphagnum* mosses and you will discover that companion vascular plants like the extra moisture and insulating properties of these mosses. Allow tendrils to touch the water to encourage the absorbing process. *Sphagnum* planted in open areas away from a direct water source will benefit from extra watering. A quick glance will tell you whether these plants are getting thirsty. Since this moss thrives in low-nitrogen conditions, it is important that you do not apply any fertilizers. In particular, calcium fertilizer can be lethal to the plants.

Sporophytes are short in stature and short lived in terms of life span. If you see the clear, watery setae and round black or brown capsules, you will be lucky. A number of *Sphagnum* species are sterile and rarely, if ever, have sporophytic stages. Researchers are investigating factors contributing to this asexual preference.

You can walk on *Sphagnum* mosses for the surreal sensation of strolling on billowy clouds, but it is probably best to locate this plant where foot traffic is minimal because even well-established colonies can tear apart more easily than other mosses. Since *Sphagnum* mosses grow quickly, I am experimenting with moss lawn and living wall possibilities by using fragmentation methods. New growth has less loft and seems to hold together pretty well.

Thuidium delicatulum

delicate fern moss, log moss, sheet moss

plants 1–4 in. tall or long; leaves 1⁄32 in. long

acidic soil

Thuidium delicatulum resembles ferns at Lilliputian scale with delicate branches shaped like fern fronds. Take out your loupe to look closely at the twice-branching pinnate plants, and you will realize that the "fronds" are actually the total moss plant with tiny leaves on the branches. *Thuidium* may look dainty, but it is quite hardy. In fact, this moss is a frequent choice for a moss lawn. In my region, this moss is evident as the understory in grass lawns anyhow. Under ideal conditions, it grows horizontally with determination, and it tends to invade other mosses. It will creep its way onto mounds of upright-growing *Dicranum* or *Leucobryum* and quickly engulf these host substrates. On more than one occasion, *Thuidium* has compromised my original moss design. Once it even happened while my moss garden was blanketed in snow. When it melted, my spiral design of *Ceratodon* and *Leucobryum* had lost the distinctive lines between species, thanks to *Thuidium*.

Depending on season and sun exposure, *Thuidium* mats can be a gorgeous green or yummy yellow. Like *Hypnum*, this moss is particularly sensitive to the amount of sunshine it gets. Where I live, this amazing plant is mainly green throughout most of the year, but when the trees drop their leaves it shifts from emerald jewel tones into golden glory in a matter of a few days. In my mossery, I moved a yellowy *Thuidium* mat from the sun section to the deep shade production area; within five days, significant green had happened, and soon the entire mat was completely green again. When it is moist, colors are intensified. The color is more opaque, similar to matte finish rather than translucent like some other moss species. When parched the leaves shrink slightly, but the moss just looks dry and is still easily identifiable. Sporophytes sport a range of colors during the maturation process, ending with a bronzy golden tint.

Like other *Thuidium* species, *T. delicatulum* is widespread throughout the world and is extremely commonplace in my locale. It has the ability to thrive on many surfaces. It grows on rotten logs as well as live trees. It can creep its way onto walls and pavements in shady, moist locations. You can plant it directly on the ground (preferably on acidic soil), but it will grow on gravel or rocks as well. This moss is a wonderful choice for just about any project and can be planted wherever you want it. It does prefer shade, but as mentioned earlier can tolerate partial sun and seasons when sheltering leaves are absent. This species thrives on lots of moisture and spreads faster with supplemental watering. Place this moss near a water feature, such as a bubbling fountain or cascading waterfall, and it will be a happy camper.

For an immediate and impressive presence, interleaf mat edges like interlocking pieces of a jigsaw puzzle. This process will encourage colonies to mesh together into one gigantic expanse of green (or maybe golden) grandeur. Alternatively, dotting fragments on bare soil

will yield desirable results within a couple of years. To tell the truth, I have been pleased with results in as little as six months, particularly if the moss is planted in the fall. Be aware that birds may make daily visits to steal *Thuidium* fragments for their nests or to snag earthworms living under the thick mats.

from concept to garden plan

DESIGNING WITH MOSSES

Inspired by a plethora of moss options for your landscape, your brain is most likely spinning with ideas already. It is easy to get carried away with notions as you dream about making mosses a reality in your garden. "I want that" attitudes need to be tempered with practicality and planning. In conceptualizing a good design, you need to consider function as well as beauty.

Before you move forward with big ideas, assess your desires in terms of intended locations and plant requirements as well as time, energy, and financial resources available to achieve your mossy goals. Once you have determined these parameters, you can proceed with incorporating the elements of any first-rate moss garden design—flow, balance, emphasis, texture, focal points, and other plants that complement the enchanting miniature scale of mosses.

When conceptualizing your moss garden, plan to enjoy it as a tranquil retreat as well as a useful outdoor living area. Your orchestration of elements in this design process will ensure an effective blend of function and landscape artistry. Keep in mind that this new space will provide you an opportunity to escape from the hustle-bustle of everyday life. Additionally, consider your perspective or view from inside your home. That way, even when inclement weather keeps you indoors, you can appreciate the lushness of your picturesque moss landscape. Last, consider curb appeal. Look at your intended area from all angles to achieve a cohesive design plan that provides garden appeal from several different vantage points. Regardless of your preference for whimsy or formal symmetry, evaluation of microclimate conditions and functional needs is an important step in the planning process.

SITE ASSESSMENT

Whether you are the proud owner of an expansive estate or have a postage stamp of a front yard, you can always find locations for intentional moss planting. Sometimes the site screams for moss solutions. The deep shade under a tree where you've been struggling for years to get something (or anything) to grow or the steep hillside that suffers from erosion issues are clearly areas that need attention. Whispers to our inner spirits can also guide the intuitive placement of moss features. Solving environmental issues and achieving aesthetic appeal require an assessment of factors that could affect the long-term success of your moss endeavors.

You can conduct your own site assessment by examining aspects that are pertinent to mosses. You will also want to be aware of any potential issues from existing environmental

PAGE 134 Choosing the appropriate moss species is key to successful designs. *Atrichum angustatum* can live in shade or sun in the worst of soil conditions.

OPPOSITE A moss lawn complements trees and shrubs in the Richmond, Virginia, garden of Norie Burnet.

factors so that you can make better decisions on where to plant mosses or how to modify the area to suit the needs of the mosses you have chosen. Some aspects of location can be changed while others you must accept and acquiesce to by adjusting your methods or changing your choice of moss species.

Sun exposure

First, I always look up to the sky. Orientation of the moss garden space to the sun's path helps you understand how much sun exposure mosses will receive in a day. Determine which direction you are facing by using a compass—either the old-fashioned kind or the new smartphone application. You can even get a phone app that tracks the sun's path during different seasons. Ascertain whether you have morning or afternoon sun and for how many hours. In the Northern Hemisphere, north-facing locations retain more moisture than south-facing areas. The idea that mosses grow on the north side of trees is true, but in my region mosses can also face west, east, and south and thus should not be relied on by Scouts to guide them out of the woods.

Be alert to places where the sun filters through the leaf canopy at specific times during the day, creating hot spots even in overall shady sites. My moss garden is really my front yard. It mainly faces west, but one section gets more afternoon sun from the unobstructed north side. Back in my shade area, which maintains this status all year long, I still have to deal with intense, focused spots of sun that occur sporadically during summer days. Even in a shady area, you may need to plant a sun-tolerant moss in identified sun pinpoints.

While still looking up, assess your leaf canopy. Notice trees in your neighbor's yard that affect your yard. Are deciduous trees (trees with leaves that fall off each year—for example, maple, oak, poplar, and dogwood) present? Based on where you live, the actual leaf season can often be shorter than the barren state. An area that is shaded in the summer by deciduous leaf cover may very well have sunshine for more of the year than not. In the mountains where I live, the initial greening of trees starts in mid-spring and the color splash is over by mid-to-late fall. For a few weeks, it seems like the leaves will never stop falling. Unlike states farther north where leaves cooperatively fall off in one fell swoop, in the South some leaves (especially oak leaves) hang on and continue to drop into the late winter and early spring.

Do not look down yet. Do you see any evergreen trees? Evergreens include conifers like pine, hemlock, fir, and spruce as well as broadleaf trees that maintain a leaf cover the entire year such as magnolia, eucalyptus, holly, and live oak. Evergreen species provide shade throughout all seasons. Mosses thrive under these species of trees because of the shared preference for acidic soil. Additionally, a bed of pine or hemlock needles provides good drainage for *Leucobryum* and *Dicranum* moss species. In climates where winter snows last for months, the fact of needle coverage versus leaf canopy is of less importance. Winter shade is not too important if mosses will be covered with a snow blanket anyhow.

Because losing your leaf canopy definitely affects the amount of sunlight on your mosses, you must be alert to potential tree diseases or pests that might affect your canopy. Within the first month after I installed a 350-foot moss oasis, two of the three pine trees (*Pinus virginiana*) that provided shade had to be cut down when attacked by the southern pine beetle. Another voracious bug, the hemlock woolly adelgid, is destroying my beloved hemlocks (*Tsuga canadensis*), leaving once-regal forests looming like gray ghosts across the mountainsides. The moral of these stories is that you should research any potential insect infestations or diseases that might threaten your trees or shrubs before planting your moss garden around them.

Speaking of shrubs—bushes, large-scale plants, and even ferns can provide shade to

TOP Your shade garden may not always be shady. My garden has spots of sun in different locations throughout the day.

This rhododendron towers like a mighty oak over the miniature landscape of mosses underneath, providing shade throughout all seasons for mosses that never die back.

Banking on mosses

"Joy, pure joy." These three words describe moss gardening for George and Carol Vickery of Brevard, North Carolina. Gardening together has always been a part of their married life. In Florida, they worked diligently to achieve a manicured and pristine landscape. In retrospect, it was not much fun. In fact, they admit to griping at each other under the scorching sun while applying the full gamut of "green thumb" chemicals, deadheading blossoms, and constantly mowing their expansive grass lawn.

Both Carol and George get giant smiles when sharing the joys of their own moss gardening experience. Originally from New Orleans, George visited these mountains regularly each summer as a boy and was intrigued with mosses. Carol, a native of St. Louis, always enjoyed mosses and ferns when hiking. With a shared dream of moving to God's country, they retired to Connestee Falls, one of the most ideal places in the world for mosses. Little did they realize that soon they would be adopting green as their favorite color in their golden years.

Having transplanted themselves into an environment where mosses abound, they decided to use moss intentionally in their new landscape. The Vickerys wanted to emphasize the natural connection to their surroundings by blurring the boundary between outdoor living space and the adjacent woods on their property. As they put on their moss-colored glasses, they could envision a mossy area on the hill above the stone patio and retaining wall. A steep clay bank 62 feet long by 12 feet tall ran behind the house, so solving erosion control issues and masking the ugliness was a functional priority. From their inside sun room, they wanted grandchildren to see their own visions of green sugar plums, not red clay. Mosses met all of these needs and desires.

A site consultation with me helped them learn which moss species were already growing on their land and which ones could be transplanted into intended locations. After this personal moss gardening lesson, the Vickerys were ready to get started. Methodically, gathering mosses from an old logging road on the edge of their property and purchasing specialty mosses from my mossery, they have transformed the patio hillside into a moss feature with moss art logs over a couple of years. The 60-to-70-degree slope behind the house is now a solid moss bank, and erosion issues have been alleviated. They used *Thuidium* colonies held down with landscape staples and small twigs until rhizoids effectively connected. George has started a special moss fairy garden for their grandchildren on the side of the house, too.

George and Carol don't work nearly as hard at moss gardening as they did at traditional gardening. Following my advice to water and walk on their mosses when possible, they have achieved successful results. An array of moss greens connects their garden spaces to the natural landscape. As Carol concludes, "It is just cool to have a moss garden."

George and Carol Vickery enjoy the magic of their evolving moss garden in Brevard, North Carolina.

LEFT The new fairy section in the Vickery garden is a favorite play place for grandchildren.

mosses. You may also have shade from privacy walls or an exterior wall of your house. On the flip side, picture windows or bright, white walls may reflect sun back into your intended location. Remember to check perimeter areas to assess the potential shade or sun factors from your neighbors' property, too.

Topography

Next, pay attention to the topography as you continue your scan of the area. Are there dips or humps in the landscape or is it flat? Do you have any steep slopes that experience soil erosion problems? Wouldn't it be great to eliminate the precarious process of weed eating on an incline? Mosses such as *Polytrichum* and *Atrichum* could be your solution. Do you have any areas where rainwater rushes through during a heavy thunderstorm? Even if you have addressed run-off issues with riprap, you may want to green the ditch with *Thuidium* and *Bryoandersonia* mosses that grow on rocks and help slow down the flow.

Low-lying areas are prone to water pooling, which causes root rot in many plants, but these soggy sites or boggy settings are ideal homes for moisture-loving mosses including *Plagiomnium* and *Climacium*. In addition to water flow or accumulation, consider atmospheric moisture, too. If you live in the high mountains, heavy

Several moss species naturally grow on rocks. This stone dam has been planted with *Ceratodon*, *Entodon*, *Hedwigia*, and *Plagiomnium*.

mists from clouds will be advantageous to all moss species. Likewise, if your property is near a river or lake, your moss garden will benefit from morning dew and low-lying fog.

Climate, weather, and microclimates

Naturally, weather patterns influence the ability of moss plants to grow and thrive. General climate factors and microclimate circumstances affect water availability, minimum and maximum temperatures, and the amount of sunshine. Check out the weather statistics for your region. What is the temperature range (daily, monthly, and yearly)? What is the annual rainfall? Does it rain every day? Does it rain more in one season than another? How much snowfall occurs? Does the snow melt or does it last for a full winter season? Differences in snowfall can influence growth versus dormancy. You do not have to be concerned about it getting too cold for mosses to survive, but sometimes stresses can trigger dormant stages. You will also want to consider the impact of seasonal snowmelt and the subsequent consequences of harmful road salt or excessive water runoff in your garden. Wind patterns affect how quickly mosses dry out, so pay attention to direction and speed of wind gusts.

Look for niches of moisture that already exist around your property. Even your driveway may have cracks or corners that accumulate rain. In your survey, do not forget to include microclimates and impervious surfaces with a high heat index, and remember that certain mosses such as *Bryum*, *Ceratodon*, and *Entodon* species enjoy the extreme heat of pavement. Hardscape elements increase nearby temperatures in gardens. Hard surfaces, including roofs, can increase amounts of surface water runoff. You can mitigate environmental issues related to stormwater if you channel this overflow into your mosses.

Mosses already present

While you are looking around, go on a moss scavenger hunt. What mosses can you find growing in your lawn? Skirting the trunks and roots of trees? Around the edges of your property? At the base of the gutter's downspout? Hovering at the air conditioning unit? In the cracks of a rock wall? In drainage ditches? Pay attention to every nook and cranny where mosses might already be growing. Identify which species are in sun or shade. Be observant of microclimate conditions where mosses have taken up residence. It is just plain common sense that these existing mosses should rank at the top of your list of promising species.

Water and power

As you continue your assessment, your site evaluation checklist should include accessibility to a water source—either an outdoor spigot from your home's tap water, rainwater collection options, or an on-site pond or stream. Additionally, having access to a power source enables the use of electric-powered garden equipment in lieu of petroleum-powered leaf blowers.

Invasive plants

You might as well make a note of any invasive plants or weeds, which are sure to present future issues as you maintain your mosses. The list of invasive plants is extensive. I suggest that you research which species are of most concern in your region. After identifying invasives, my advice is to *get rid of them*. Unfortunately, you will find that mature plants may be difficult to eradicate. The next bit of advice: *Avoid purchasing any aggressive species*. You will find that nurseries and garden centers offer all kinds of invasives and weeds with attractive blooms and foliage. Years ago, I made the mistake of planting periwinkle (*Vinca minor*) for hillside stabilization (and because I liked the blue-purple flowers). Now I have the annoying task of having to pull this invasive ground cover out

of my mosses. Another popular plant used to line flower beds is monkey grass or lily turf (*Liriope muscari*). While not formally categorized as invasive, it is one of hardest plants to dig out and seems determined to persist.

While some plants are universally invasive (such as North America's popular English ivy, *Hedera helix*), annoying plants may vary from region to region, state to state, country to country. Unfortunately, many of us have inherited plants historically introduced as preferred landscape plants or intentionally planted as agricultural solutions that have proven to be invasive. Renowned American landscape architect Frederick Law Olmsted recognized the disruption of natural ecosystems by invasive plants and warned of this problem in 1865 in a report about Yosemite Valley. Yet, some of the plant species Olmsted recommended in his designs for parks across the country have proven to be invasive. English ivy, privet, Japanese honeysuckle, bittersweet, and multiflora roses planted in Iroquois Park in Louisville, Kentucky, in Central Park in New York City, and on the Biltmore Estate in Asheville, North Carolina, are impacting those landscapes and have escaped into nearby woodlands, fields, and roadsides. Luckily, I inherited my grandmother's attitude about ivy: she thought it was snaky and avoided planting this decorative vine. Still, I deal with ivy from the perimeter woods aggressively sneaking into my mosses.

Be advised: Although native plants are never categorized as invasive, some species may have aggressive tendencies. Once again, I discovered this issue the hard way. Finding mountain bluets (*Houstonia caerulea*) growing among mosses along streams, I thought this ground cover with tiny blue flowers would make a nice complementary plant. Wrong. While attractive, intentionally introduced bluets have invaded my mosses and spread to new areas. They are almost impossible to remove because the bluets' tiny roots interconnect within moss colonies, so I am now resorting to digging out whole sections of mosses smothered by native bluets.

What invasive plants are common to your region? If you want to be a responsible land steward, avoid using invasive plants in the first place. As a moss gardener, you will not appreciate the aggressive growth of invasive plants that could overtake your mosses. To learn more about invasives and weeds common to your locale, visit Web sites sponsored by land grant universities, agricultural extension offices, and native plant societies.

Beware of aggressive natives like bluets. Tiny leaves and blue flowers look great in spring, but this ground cover will thoroughly invade your mosses.

DESIGN TIME

When you have done a thorough assessment of your own property and gained an awareness of the factors that could impact your moss gardening success, it is design time. A formal landscape design drawn by a professional is great, but you can sketch out a rough plan for your own purposes. Experienced gardeners may elect to visualize the intended space in their heads without the need to put ideas on paper. For planning purposes, creating a shopping list of mosses, complementary vascular plants, and supplies will be valuable to keep yourself organized and within budget. As you envision your moss garden, plan to solve environmental problems and to create tranquility at the same time.

Form follows function

As you synthesize your creative ideas into a cohesive design plan, keep in mind that form follows function—your design concept and resulting moss efforts should serve the basic purpose of enjoyment. In the orchestration of elements, make sure to provide an entrance, strolling paths, and opportunities to sit and revel in the sublime serenity of your moss garden. Your entry should beckon visitors to come in and experience the magic of your moss retreat. Paths and walkways could guide visitors toward points of interest or pocket gardens that you want to showcase. The meandering route can be composed totally of mosses that tolerate moderate foot traffic.

At Edenwoods, Norie Burnet's moss garden in Virginia, marvelous moss pathways lead visitors to various features. In Kenilworth, at the home of John Cram and Matt Chambers, more than a mile of mossy trails wind through exceptional gardens that display sculptures crafted by some of the nation's leading artisans. In contrast, Dale Sievert's garden in Wisconsin includes paths of wood chips and mulch. His paths lead toward enticing destinations—numerous garden beds of moss mounds and carpets, hosts of hostas, bunches of brilliant-colored flowers, and water features reflecting an Asian influence.

Footpaths, stepping-stones, and benches exemplify the functional aspects of my moss garden. With the help of my son Carson, I added a pebble footpath and stepping-stones to lead people in different directions. Benches and seating areas offer the opportunity to relax and contemplate life. Ironically, I have to coax garden visitors to take off their shoes and leave the path to fully experience the moss carpet. Today, mosses are sneaking into the gravel corridor. Over the years, I've casually tossed moss fragments from the roots of removed weeds into the gravel. Now the path is naturalized and mossified.

Poplar log disks (cross-cut pieces of a log called cookies) acquired from a lumberjack competition lead garden guests through the arched arbor entrance made of lichen-covered mountain laurel branches. I learned two valuable lessons from this trial footpath. First, during the protonema stage (early moss growth), the cookies got really slippery (especially where I tried the moss milkshake method). Now that mosses are effectively covering my cookies, the danger of accidentally slip-sliding away has passed. The second lesson learned was quite a surprise to me. I observed that faster growth occurred on the cookie that was painted purple (with high gloss enamel paint, no less) than on the rough-cut natural log surfaces!

Your design plan might include hardscape elements for crafting a useful and charming outdoor living space. Patios made of stone, brick, or concrete pavers offer a platform for enjoying life in the open. Boulders can add larger scale and drama as well as serve as seating. Stone or brick walls can provide division and emphasis. Short walls offer extra places to sit, while tall walls provide privacy. If you include a firepit, use good sense and avoid planting mosses around it lest fire sparks escape. Water features

TOP Become a moss artist by combining species into innovative designs.

At Norie Burnet's garden in Virginia, moss and stone pathways lead to features such as this small stone pond.

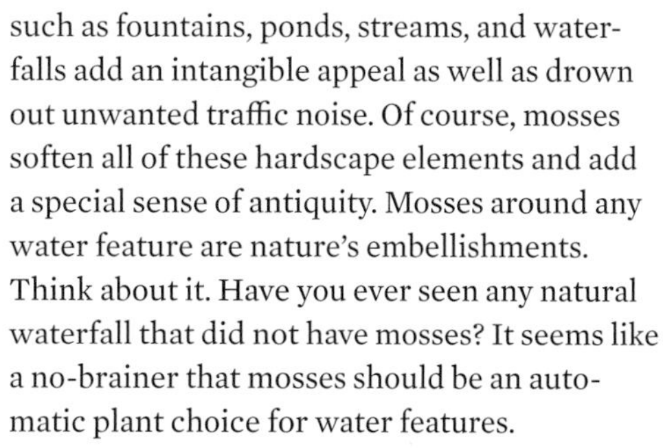

such as fountains, ponds, streams, and waterfalls add an intangible appeal as well as drown out unwanted traffic noise. Of course, mosses soften all of these hardscape elements and add a special sense of antiquity. Mosses around any water feature are nature's embellishments. Think about it. Have you ever seen any natural waterfall that did not have mosses? It seems like a no-brainer that mosses should be an automatic plant choice for water features.

"Sit down and stay a spell" is the next order of business. Yes, sitting is an important functional aspect. Foster enjoyment in your new moss garden by strategically placing benches, boulders, or outdoor furniture where you can pause to reflect. Besides the solid stone bench bestowed as a gift from my neighbor Tommy, my garden seating includes benches made from moss-covered laurel and rhododendron logs, moss art stumps, and my beloved purple rocking chair. My favorite footstool is composed of cushiony mounds of *Dicranum* moss. Low benches are nestled under the shade of a Japanese maple tree and in an alcove of rhododendron for an intimate experience with nature.

Set the stage to enjoy your moss garden at night by adding dramatic spotlights that accentuate focal features. Use flood lamps and spotlights to emphasize the undulations of your moss landscape. Additionally, you may want to install hard-wired lighting, battery-operated LEDs, or solar-powered lanterns to line your pathways for an evening stroll. If you want to follow a traditional Japanese approach, use stone lanterns for exceptional garden illumination. For an out-of-this-world experience, try using glow stones for paths. These solar collectors (plastic or glass) illuminate at night to light your way. If you venture into your moss garden under a full moon, I guarantee you will be astonished at the high definition of every mound and ridge.

Dale Sievert's garden includes mulch paths with moss features and a host of hosta varieties, creating intrigue at every turn.

Design with moss in mind

Since 1969, Dale Sievert of Waukesha, Wisconsin, has been transforming his barren, treeless property into an acre of exquisite gardens. Formal brick walkways encase symmetrical flower beds in the front. In the backyard, natural paths lined with more than a hundred species of hostas and thirty-plus species of mosses lead to an array of terraces and garden destinations including water features and his beloved Japanese garden. Beyond incorporating mosses into his garden landscape, Dale wears the crown as the king of moss container gardening. Can you imagine more than 250 pots of mosses strategically placed all around his fourteen different gardens?

Dale's early garden design was stimulated by a visit to Bellingrath Gardens in Mobile, Alabama. The oval-shaped Great Lawn made a strong impression. Later, a trip to Virginia prompted him to create his own Williamsburg garden section. Without a written plan or blueprint, Dale synthesizes the best elements, explaining, "I can see things in my mind the way they should be."

In the summer of 2005, Dale visited Foxfire Botanical Gardens in central Wisconsin, where the small Japanese garden inspired him. He was enthralled with the light filtering through deciduous trees onto large boulders skirted with moss colonies. Not long after, Dale began intentionally introducing mosses and started his own Japanese-style garden. He used mosses found nearby and mimicked growing conditions. Rather than waiting for mosses to grow in, he planted colonies close to each other to obtain instant gratification.

From 2007 onward, Dale has continued to introduce mosses and expand moss presence in his garden features. Since he returned from a trip to Japan, he's been busy again creating a new water feature with bamboo and his preferred moss species, *Anomodon rostratus*, which grows abundantly in his limestone-based soil. New growth has a vibrant yellow-green color. In mature colonies, *A. rostratus* tends to mound up, resulting in the rolling look Dale prefers over the flat effect of a grass lawn.

Retired from his career as a technical writer and engineer, Dale works tirelessly at creating and maintaining his gardens by himself. He patrols at least once a week, plucking out weeds as soon as he sees them. He notes, "I have to be vigilant and diligent in the removal of litter and debris. Otherwise, in no time at all, the area will look tacky." Relying on natural rainfall, Dale accepts dry and dormant stages, particularly in the summer. He provides supplemental watering only as a "perk up" measure before visitors tour his gardens. He is concerned that the calcium and magnesium ions in the well water he uses will eventually cause suffering or death to his moss colonies, but he does water his moss containers regularly.

When it comes to his mosses, Dale laments, "I regret not appreciating moss for the first sixty-three years of my life. I am now seventy-one. In my next life, I will start from infancy."

Stones are arranged to emphasize contrasting mosses.

OPPOSITE With more than 250 moss containers, Dale Sievert is the moss container king as well as the weeding police chief.

TOP Moss-and-stone patios have great appeal.

Mosses soften these stone steps at the Kenilworth moss garden and add a sense of antiquity.

Right plant, right place

Many successful gardeners follow the rule "right plant, right place," which in many ways is a reaffirmation of "form follows function." As moss gardeners, we will adopt this guideline, too. Identifying micro-niches of moisture in your garden and locations conducive to moss growth are central considerations in choosing the right place. Knowing which moss species prefer only shade, partial shade, or direct sun will help you determine the right plant, or in our case, the right moss. In truth, it is easiest to start your moss gardening journey in a spot where few challenges exist, so my advice to novice moss gardeners is to choose a shady location.

Many moss gardeners report better success when planting mosses under a conifer cover, but other moss gardeners avoid trees that continually drop needles. Needle litter can be quite annoying. Just because evergreens keep foliage all the time, do not assume that they do not shed needles or leaves. Probably 95 percent of the golden pine needles that cover my moss carpet in the fall are from my neighbors' yards. Although not as problematic, minor needle shedding happens throughout other months, too.

You might as well face the fact that if mosses are located under any kind of tree, maintenance will be necessary. Under deciduous trees, you may have to deal with leaf litter longer than you want. I can blow my entire moss yard clear one day, and the next day dead brown leaves cover my green carpet again. If you plant your moss garden on a windy hillside, maybe most of the leaves or needles will blow away and reduce your chores.

Be informed: While mosses grow well under fruit and nut trees, you may face messy nut litter after squirrels stop in for a tasty treat. Still, mosses may be the right plant under all types of nut trees. I have seen *Entodon*, *Hedwigia*, and *Thuidium* mosses growing under walnut trees where no other plants ever grow.

What better place for a moss garden than at the foot of a waterfall? This ideal spot is located at Camp Merrie Woode in Sapphire, North Carolina.

TOP If you have only a small space, consider using planters or containers for moss features.

If you plant your mosses under any kind of tree, you will need to periodically remove needles, leaves, and debris.

Do not forget that soil pH can be a factor in choosing the right place for your mosses. Since conifers and nut trees are known to love acidic soil, they provide a desired environment for many moss species. If you do not have any pine trees or majestic oaks, you may need to modify soil conditions and choose sun-tolerant moss types to obey the "right moss" portion of this gardener's adage. Bryophyte-specific planting methods outlined later in the book will elevate your efforts in achieving "right moss, right place."

Companion plant choices

Your choice of companion plants for your mosses needs to match the environmental setting of your garden with the typical considerations of sun exposure, soil conditions, and seasonal interest. All types of native plant species as well as ornamental perennials and annuals will complement your mosses.

If you are planting your moss garden in the shade, pick shade-tolerant companion plants like hostas and ferns for their impressive foliage. Perennial or annual flowering species can add a pop of seasonal color if you have sections that are sunny. As much as I love mosses, I do enjoy early spring bursts of intense yellow from my daffodils—a fond reminder of my mother's garden. With my purple obsession, I tend to plant perennial and/or annual flowers such as lavender phlox, magenta impatiens, and maroon dianthus to serve as summer borders to emphasize moss sections. The splendid leaf color of my Japanese maples is impressive, most exceptionally in the fall. Evergreens with leaves, such as rhododendrons, along with conifers like blue spruce trees, could complete the winter picture. Throughout all seasons, mosses provide a constant green backdrop to emphasize the appeal of other garden plants, bushes, and trees.

In creating a moss garden feature, you can emphasize the miniature aspects by choosing companion plants that maintain a small stature. Beguile garden visitors into taking a closer look at your distinctive mosses by the inclusion of other small-scale plants—tiny, bright-colored flowers, delicate ferns, dwarf conifers, and petite trees. This year, I am adding dainty tête-à-tête daffodils for even more miniaturization. In this case, mosses are the stars and other plants either complement or accent them in supporting roles.

Some of my favorite miniature plants are native. I am particularly fond of the white-veined leaves of downy rattlesnake plantain (*Goodyera pubescens*), an orchid native to North Carolina, which has a shallow, horizontal root system. For more impressive year-round foliage, I've planted galax (*Galax urceolata*) and Oconee bells (*Shortia galacifolia*), which both have shiny roundish leaves. Like evergreens they keep their leaves all year, but in winter they shift into crimson shades. Other personal preferences include small-scale ferns such as ebony spleenwort (*Asplenium platyneuron*) and Appalachian polypody (*Polypodium appalachianum*). With white flowers in spring, red berries in fall, and consistently dark green leaves, partridge

Creeping, prostrate partridge berry (*Mitchella repens*) is a lovely complement to *Hypnum imponens*.

CLOCKWISE FROM TOP LEFT
Endangered and quite rare, Oconee bells (*Shortia galacifolia*) are another miniature plant to complement mosses. The leaves last through the winter, changing from green to burnished red, and a delicate blossom appears in the spring.

One of my favorite miniature plants to complement the diminutive aspects of mosses is a wild orchid, downy rattlesnake plantain (*Goodyera pubescens*).

Small Appalachian polypody (*Polypodium appalachianum*) ferns complement *Dicranum scoparium* moss mounds.

Benefits of native plants

Since you are going native with mosses, why not consider going native all the way? Native plant enthusiasts point out advantages to gardening with plants that have evolved to live in your locale. Like mosses, native plants and wildflowers require no fertilizers, pesticides, or herbicides, making them environmentally friendly. Other advantages include water conservation, lower maintenance, and compatibility with the ecosystems of surrounding natural areas. In developing your plant list to complement mosses, plan to create a harmonious ecosystem. Keep in mind that plant communities include symbiotic relationships or ways in which one species helps another in some way. Choose native plants and wildflowers accordingly.

Utilizing landscape and garden design principles, plan a place not just for human enjoyment but for wildlife, too. Native wildlife—birds, beneficial insects, and all types of interesting little critters—prefers native plants. Decorative birdbaths and shallow vessels offer an occasional splash bath or drink for wildlife visitors. The cool and moist environment of mosses provides a preferred micro-habitat for all types of wild guests. It is always a delight when honey bees and butterflies drop in to chill out in my moss garden.

To learn more about native plants and wildlife found in your region, check out information provided on the Web by native plant societies, wildlife organizations, and government agencies. Rest assured you will find lots of useful information about vascular plants and indigenous wildlife. Ironically, you may be disappointed to find that mosses (bryophytes) are overlooked in lists of recommended native plant species.

berry (*Mitchella repens*) offers a striking contrast to my mosses. My absolute darlings are the native dwarf crested iris (*Iris cristata*). These purple irises, with tinges of yellow, are perfect miniatures in my fairy garden.

Almost every wildflower complements mosses. Trillium (*Trillium grandiflorum*) and jack-in-the-pulpit (*Arisaema triphyllum*) offer the extra advantage of providing shade to the mosses growing underneath their wide leaves. In bog gardens, carnivorous plants such as pitcher plants (*Sarracenia* species) thrive in colonies of *Sphagnum* mosses.

Spiritual elements

Inspired by the grand temple moss gardens of the Far East, many people desire to create a spiritual space. Along with mosses, other essential components resonate from ancient gardens. Water features such as ponds, streams, and waterfalls are key elements representing real or mythical lakes. In dry gardens, raked areas of white gravel and sand symbolize water. Manmade hills or mounds denote mountains. In larger gardens, water surrounds islands complemented by arching footbridges. Stone lanterns and water basins are paired together as essential components of tea gardens. Water cascades in bamboo fountains from one hollow stem to another. As Westerners attempt to emulate spiritual elements, statues of Buddha have become a common component in Japanese-inspired gardens in America.

Follow your own heart. To honor European ancestors, you might carry on the ancient tradition of stacking stones to create small pillars. Historically considered sacred, such cairns exist all over the world. The point is that you do not have to be a Zen Buddhist to experience a sense of spiritual renewal and connection to nature in a moss sanctuary. You can place symbols of your own religion in a moss garden, or none at all, to create a retreat for your spirit.

TOP You can use mosses to create a place of spiritual renewal and meditation in your backyard.

Expect natural transitions to occur. This volunteer rhododendron sprouted at the base of a decorative stump in my moss garden—a gift from Mother Nature.

Artistic touches

Moss gardens offer an effective backdrop for showcasing garden art and sculpture. In creating points of interest with garden art, let your own personality shine. My connection to nature is obvious to those who visit my moss garden. When they enter my moss retreat, their eyes are drawn to locations that display impressive moss art transformed from forest driftwood—knotholes, stumps, and hollow logs—ranging in size from a few inches to several feet across. To my delight, a volunteer rhododendron has arisen from an apple log swathed in *Hypnum imponens*. This particular naturalized moss art reflects the role of mosses in the forest life cycle as a nurturing environment in which seeds can germinate. Naturally, moss art logs eventually rot away (unless you coat them in marine resin, which ruins the natural effect for me).

Add elements of serendipitous surprise and whimsy in your moss garden design. While I am not a fan of garden gnomes, I do like concrete bunnies. Mine are positioned to peek around rocks. After several years and numerous failed attempts with the moss milkshake method, mosses have finally self-introduced on my bunnies. I have found that the more porous the statuette, the more likely it is that mosses will start to creep onto the surface. I have also arranged elongated smooth rocks 3 to 8 inches long (what I call hot-dog rocks) on the ground to resemble daisies or coneflowers. I've interspersed mosses between the stones to emphasize them and added purple coneflowers (*Echinacea purpurea*) that tower above my moss and stone flowers.

In thinking about designing with rocks, think back on function again. Along the street, the edge of my moss yard kept getting damaged. I decided to use hot-dog rocks in a freeform design with scattered moss fragments. Now cars can park or drive over this section without creating muddy ruts—far more functional, and whimsical, too. For further inspiration, check out the natural art forms created by the internationally acclaimed environmentalist and sculptor Andy Goldsworthy.

Add whimsy with a creative arrangement of stones and mosses together.

RIGHT Fairies are at home in Mary Beth O'Connor's magical garden with a landscape of miniature trees and mosses.

I created my fairy garden one February night under the full moon.

Fantasy fairylands

If you are in touch with your inner child or have real children in your life, set aside a small area in your moss garden for a fantasy fairyland. Magical mosses and flitting fairies just seem to go together, and they can spur creative thinking and imagination. Like numerous other people, you may have a fond memory of playing in the mosses, creating rooms with archways fashioned from twigs and furniture made of toadstools (mushrooms). As you recall the experience, you may still be able to feel the softness of the cushiony green carpet and the refreshing coolness experienced while playing for hours in your own world. Pass this pleasure on to your own children or grandchildren by encouraging them to help with the process of planning and creating this special corner of your moss garden. Let them continue the expansion of their moss fairy garden on subsequent visits. The special memories you will make together will be precious and truly may last a lifetime.

You can add all kinds of fairy elements to your moss feature that will stimulate the imaginative process. Exquisite fairy statues with flowing hair and delicate wings are available. An array of miniature furniture, footbridges, and even fairy pets is offered for sale on the Internet and at high-end garden centers. Given the popularity of fairy gardens, you will be thrilled with the selection you might find. During the holiday season, you can find some good deals on items at discount chain stores. Inexpensive, iridescent blue glass pebbles are great for representing fairy swimming holes. Introduce fun and science by adding decorative objects found in natural settings (such as quartz crystals, acorns, lichen sticks, and pinecones). Moss art logs and stumps with fairy doors will add yet another element of delight.

Build it and they will come. If you create a magical moss garden, maybe real fairies will appear. Who knows? You may be surprised to see a flickering light darting around your moss garden at midnight.

A magical fairyland moss garden

Ann Gordon's moss memories take her back to childhood, when she played quietly among tree roots making homes for the fairies with cushiony mounds of light green and creeping carpets of fern moss. She didn't need names then, and even though she has learned their botanical names, she's fine with knowing mosses just by feel, look, and habit. She has come full circle. Today, this talented author and illustrator of children's books plays, writes, draws, and lets her imagination soar in a magical fairyland in Efland, North Carolina, that she fondly calls her Mossy Hillock. Drawn to this location during walks with her loyal dog companions, Ann would rest in the cool shade of a massive oak. This grown-up fairy couldn't resist clearing debris and weeding a bit to reveal green patches of mosses.

Ann admits that rarely did she intentionally plant any moss; rather, she would empty her pockets of special gems retrieved here and there from her strolls. Over the years, she cast moss sprigs around and waited patiently for them to grow and interweave. Each day she brought a small bucket of pebbles to mark the paths up her miniature mountains, and gradually Fairy Hillock #1 evolved. From a few dwellings, her fairy moss garden has grown into an entire village of magical places, including the Crystal Palace and the Castle of the One-Winged Bird. Neighborhood children have enhanced the pebble markings leading from one mystical spot to another with a sprinkling of marbles. Ann says, "The fantastical haven is home to brownies, sprites, elves, and a host of other fairy-like entities." As she shares her magic, this brown-eyed former teacher with salt-and-pepper curly hair emits the aura of a fairy godmother.

On misty mornings when mosses are glimmering and dew is sparkling on spore capsules, Ann hears moss whispers—lessons to be learned about life and our connection with nature and the universe. This whimsical fairyland is her own special place on earth for rejuvenation of body and mind. Suffering from long-term effects of Lyme disease, Ann turns to her moss buddies to remind her of the resiliency of life. Watching mosses spring back to lushness after a rain lifts her spirits like a soothing elixir.

She calls herself Ann, the Moss Tender, and her grandchildren call her Mossy. For me, Ann is my moss *mei mei,* Chinese words for "little sister." She nurtures mosses. She pulls weeds and removes litter to encourage their growth. "Mosses don't necessarily grow where you intend them to be, they grow where they are tended," she explains.

Ann's mossy hillock is only one of her moss ventures. As I write this book, she is creating a stone and moss hopscotch area. And, as always, her mind is spinning with stories whose main characters frolic in the mossy nooks and mysterious crannies of her hillock heaven. As she brings all this magic to life through words and pictures, she illustrates her conviction that mosses are a life-giving force for all of us.

Ann Gordon's mossy hillock in Efland, North Carolina, is ready for fairy visits.

FINANCIAL CONSIDERATIONS

As you've been going through the site assessment and design aspects of planning your moss garden project, you may have already started to hear the cha-ching of potential costs ringing up. Now is the time to do a cost analysis of the moss design you have dreamed up. The most crucial financial consideration is the scope and size of your moss endeavors. How many square feet can you afford to buy, install, and maintain? Be realistic and plan your garden within your means.

In calculating costs, consider professional advice, labor (site preparation, installation, and ongoing maintenance), supplies, elements for enhancing outdoor space, complementary plants, and of course, mosses. Live mosses for sale, which are shipped fresh in containers or as prevegetated mats or living walls, are priced by the square foot. As a general rule, sun-tolerant mosses cost more than shade-loving moss species. Specialty or rare moss species are more expensive than common mosses. In contrast, dry mosses, which are shipped in boxes with limited species options, are sold by weight or volume at cheaper rates.

Of equal significance, how soon do you want to enjoy your moss garden in its full glory? If you want instant gratification, whether you hire professionals or do it yourself, you need to consider the monetary aspects of having it happen quickly. Contiguous planting of moss expanses can be expensive; the prices of most moss garden projects installed by professionals are calculated by the square foot. Even if you wait for mosses to grow in, there will be costs related to site preparation, garden maintenance, and acquisition of any additional mosses.

It stands to reason that if you involve professionals or experts, your costs will correspondingly increase. To ensure a cohesive design plan, you may want to engage the services of a landscape design architect. Professional landscape contractors have the skills, equipment, and work crews to lighten your load in terms of site preparation (particularly moving boulders, controlling erosion, and addressing other hardscape issues). If you are installing a stone patio or wall, an experienced mason can breeze through this task. If you want to install an irrigation system, hiring irrigation specialists is essential in designating zones and gauging appropriate water pressure for the system.

Yet even experienced and knowledgeable professionals may not possess insights into the specific needs of mosses, so it could be beneficial to include a moss gardening expert on your team. Involving a moss specialist early in the process can ensure realistic planning and execution of your moss installation. Otherwise, you may end up stymied by elaborate designs that may not be logistically feasible within your budget parameters. For landscape architects, delaying the participation of a moss specialist can result in change orders and additional unexpected costs.

After adding up all the possible costs, you may need to scale back your design or plan your moss garden in phases to stay within your budget. To save money, you can address site prep, acquisition of mosses, installation, and ongoing maintenance on your own. The do-it-yourself method will increase your time and work load but it is a way to cut down on your direct costs. You will need to determine what your time is worth in comparison with the value of a professional turnkey installation.

DIY terrarium

If you live in the city without your own green space, you can bring a bit of green into your life and establish a connection with nature through the charm of a moss terrarium. Live mosses can thrive for years in the self-sustaining environment of a terrarium. The Wardian case, a closed protective container for plants, started the terrarium craze in the Victorian era. In the 1960s, when I was a young girl, making your own terrarium gained widespread popularity. And once again, in the twenty-first century, there is renewed interest in the art of making a terrarium. Consider your design elements and visualize the finished terrarium before following these steps to put it together:

1. Acquire a closed-lid glass container from a garden center, florist, crafts supplier, or thrift store. Clean with glass cleaner inside and out, rewipe the inside surfaces with a water-soaked cloth or paper towel, and wipe dry.
2. Carefully place a layer of small stones in the bottom of the container for drainage.
3. Cut a small section of black landscape fabric or felt to place over the drainage stones in order to prevent soil from seeping in.
4. Put a ½-inch layer of aquarium charcoal on top of the fabric layer to prevent odors.
5. Add soil—commercial potted soil mix, decomposed leaf litter, red clay, or rich loam—in a layer up to 3 inches deep.
6. If you are using other plants, place them in position and add a bit more soil around each plant.
7. Add upright growing mosses with sideways rhizoids, such as *Climacium* and *Rhodobryum*, and cover the rhizoids with a thin layer of soil. Make sure that upright "trees" or "green florets" are vertical, not horizontal. *Sphagnum* moss can be used as a growing medium or featured as another beautiful moss.
8. Next, select positions for acrocarpous (mounding) mosses such as *Dicranum*, *Leucobryum*, or *Aulacomnium*.
9. Add *Hypnum* and *Thuidium* carpet mosses (pleurocarpous species) to fill in around other plants and mosses. Interleaf edges.
10. Finally, place anchor rocks, lichen sticks, and any other decorative elements (such as fairies and miniature figurines).
11. Mist thoroughly, but not so much that pooling occurs, and put the lid on.

Locate your terrarium near a window where it receives either morning or afternoon sunlight. Each day, the inside of your closed-lid terrarium should bead up with water. Mosses should be slightly moist, not soggy. When water stops beading on the glass interior, it is time to mist again—usually about once a month. If you notice any signs of mold, take off the lid, remove the moldy portion, pour off excess water while holding your moss scape in place with your hand, and let the terrarium dry out for a few days. If you have used an open container, you will need to mist it every day.

ABOVE This impressive terrarium in a Wardian case was created by New Yorker Michael DeMeo, moss and terrarium enthusiast.

planting and propagation

ESTABLISHING YOUR MOSS GARDEN

Now that you are familiar with how unique botanical characteristics of mosses will impact growth patterns and have some ideas in your head (or planned on paper) for transforming your landscape with mosses, it is time to gather the few tools and supplies necessary to move forward. You will find a number of moss planting methods described in this chapter. Choose the one that fits your budget and timeline. By taking recommended actions, you too can make moss magic happen in your garden.

WHEN TO PLANT

In comparing planting a moss garden to any other garden, one of the biggest differences is that it is so much easier. Besides rarely needing to dig any holes, which will thrill people who for one reason or another want to avoid this activity, you are not restricted to typical planting seasons. Since mosses have their own antifreeze and not only survive but also can grow in cold weather, they can be planted in all seasons. I can testify to successful results achieved in every single month of the year. I have even triumphed in planting mosses on frozen ground, although mosses growing in colder regions where snow stays for months may experience a dormant stage and require a brief recovery period upon thawing in the spring.

The bottom line is that you can plant mosses whenever it is convenient and comfortable for you. If you get the urge to garden in the winter, and the weather provides a window of opportunity, go right ahead. If you want to follow normal garden routines, plant your mosses in the spring. If summer suits you better, that's okay too, although higher temperatures may necessitate more attention to moisture needs. Fall is a great time because temperatures typically are cooler and mosses experience less heat stress. During autumn and winter, you can proceed with fragmentation methods, which are visibly less obtrusive through the "off season." I will concede that winter planting can be chilly. Gardening in the cold is certainly a test of your Arctic survival skills. However, once you've caught moss fever, you might find yourself bundling up in warm clothing to enjoy a bit of creative moss cultivation.

In terms of what time of day to plant, once again you have flexibility with mosses. It is a matter of when you enjoy it most. Early morning is good because temperatures tend to

PAGE 162 Some species such as this *Bryoandersonia illecebra* can live on soil or rocks.

OPPOSITE When Nashville, Tennessee, photographer J. Paul Moore noticed mosses encroaching on his grass lawn, he decided to embrace the concept of a moss lawn instead. Several years later, his moss lawn is thriving.

be cooler and the dew sparkling around you is magical. Alternatively, why not plant under the twinkle of the stars? From a functional point of view, in summer months it is usually cooler at night than under the glaring sun. Whatever time of day (or night) you plant, you will want to provide supplemental watering during planting. As you have learned, higher moisture and humidity levels are desirable conditions for moss beauty and health, so you can plant in the rain or after a shower when the ground is moist. If you plant right before a heavy storm, moss installations will be put to the test right from the beginning. If newly planted mosses can handle driving rain and high winds, you can feel confident as they face other extreme weather conditions.

TOP Only a few tools are really necessary—a heavy-tined rake, a trowel, a whisk broom, and a toothbrush.

A toothbrush comes in handy for preening tiny mosses like *Ceratodon* and *Bryum* species.

TOOLS YOU WILL NEED

Moss gardening methods require fewer tools and equipment than traditional gardening. Once your hardscape is in order, you will need only a few items for planting and taking care of your mosses. I recommend a trowel, a three-pronged hand rake (or cultivator), a push-pull scraper, a small shovel, a broom and dustpan, and an electric leaf blower. You will have no need for most weed removal tools since you are going to get rid of weeds before roots are very big, right? I do keep a cheap steak knife available to use for careful removal of stubborn weeds. At the suggestion of one of my co-workers, I have dedicated a pair of tweezers for meticulous weed extraction.

While I have a bamboo rake, it requires quite a light touch to use it without disturbing mosses. It is best to avoid raking except during the planting stage when you may want to use a heavy multipronged rake. A regular kitchen broom with stiff straw works well to sweep mosses clean. Mostly I use a child's broom or my thirty-year-old kitchen broom, handmade from an apple branch. For tedious planting of stone patios, I always use a small whisk (handheld) broom and dustpan to keep my area clean and tidy. I take advantage of nature's tools such as small pointy rocks and sturdy sticks for digging out crevices. I keep a stash of dead azalea twigs (by-products of pruning) for securing pleurocarpous mosses in place, especially when planted on a steep bank. Another indispensable tool in my repertoire is a toothbrush used to clean grit from colonies of smaller moss species. As a dentist's daughter, I have always found good ways to reuse old toothbrushes.

It is essential to acquire a good garden hose. Do not skimp on this purchase. A good hose that does not tangle is worth its weight in gold.

Avoid the headaches and frustrations encountered with inexpensive hoses. While I prefer a heavy duty, never-kink style, you may consider newer lightweight, retractable hoses. All kinds of nozzles and sprinklers are available to use with your new hose. I like lightweight plastic nozzles with a choice of spray options—particularly mist, flat, and shower settings. After trying many types of overhead sprinklers, I prefer inexpensive ones that oscillate back and forth. For small areas, a bubbler style may be your sprinkler of choice. For coverage of larger areas, a pulsating machine-gun-style sprinkler may be of value. In summary, I recommend that you spend money on good hoses but go cheap on spray nozzles and sprinklers.

Comfort when planting is of primary concern to me, so I often sit directly on the mosses, oblivious to my wet posterior. However, if you want to stay dry, I suggest using a waterproof cushion, a rolling stool, or garden knee pads. The main clothing consideration is to wear flat shoes or flip-flops when planting. Do not wear clogs, shoes, or boots with heavy tread soles. Of course, you can do barefoot moss gardening anytime (that is, unless frostbite is a risk). As for gloves, it is your preference (I rarely use gloves), but form-fitting gloves of thin latex are recommended. Thick garden gloves inhibit your ability to feel horizontal surface roots and compromise your efforts to carefully remove weeds from underneath mosses.

OBTAINING MOSSES

If you want to intentionally introduce mosses into your garden rather than waiting for them to self-introduce, you will need to acquire mosses for your project. If you are lucky enough to already have mosses growing on your property, consider moving colonies from perimeter or fringe areas to your intended moss focal feature. If you have neighbors who do not appreciate mosses, ask them if you can harvest from their land. Seek out additional opportunities by networking with private property owners who have extensive woodlands. Using this scavenger approach, you may spot other potential rescue locations around you. For instance, in cities, property intended for commercial development or abandoned industrial sites may have highly desirable moss species that are subject to total destruction when the bulldozers arrive. If you notice signs of paving crews, see if you can rescue mosses from parking lot corners and along the edges of those concrete tire bumpers before resurfacing starts. If you live in the country, look for signs of logging, road construction, or new housing developments as possible moss rescue sites. Wherever you wish to save mosses, I suggest you gain permission from the owner, real estate agent, or officials on-site.

Mosses can often be reclaimed from logging sites. Of course, the idea is to rescue mosses before loggers arrive.

FROM TOP TO BOTTOM
Gently remove pleurocarpous mosses by pulling the edges loose from the substrate.

Not many tools are needed for rescuing mosses. Acrocarpous mosses might require a small shovel, trowel, or scraping tool. Even a wide putty knife or kitchen spatula can be used.

At a local greenhouse, mosses covered the substrate. My friend Ann Gordon helped me rescue *Marchantia* (liverwort) and *Pohlia* (moss) just in time before they were thrown away.

I have seen all types of kitchen utensils used in moss harvesting, including long spatulas, knives, and even pizza paddles, but you can remove most pleurocarpous mosses by hand. Gently start to peel up the moss carpet around the edges. Stick one hand underneath the colony and sweep it back and forth to help loosen rhizoids. Using both hands, grab the colony by its edges and delicately remove it from its substrate. While it can be beneficial to include original soil with the colony, it certainly is not necessary. It depends on the species. You don't have much choice when it comes to most acrocarpous mosses, but I rarely take soil with pleurocarpous growers. I use a similar horizontal retrieval process for certain acrocarpous types that have short rhizoids embedded in the soil. Insert your fingers under the colony and gently feel around to the depth where the moss colony (with its own soil) releases from the ground. After you get it started, pull up large sections by hand or use a flat shovel or machete to help reach farther underneath for even bigger areas. Sometimes you can pluck up mosses

Where to gather mosses—and where not

There are important distinctions among rescuing, harvesting, and stealing mosses that you should be aware of as you think about where to obtain mosses for your garden.

Rescuing mosses means removing them from places where they will inevitably be destroyed, such as where new residential or commercial development is slated to occur. Keep your eyes open for signs of imminent destruction; when the bulldozers arrive, it is usually too late. "For sale" signs (particularly at sites designated for commercial development or new residential communities) signal moss rescue possibilities. On many occasions, upon noticing a new roof about to happen, I have brazenly asked permission to rescue mosses before the roofers started. Just ask; usually the answer will be yes. In such situations, you might as well take all you can get because the mosses are on death row anyhow. For a group experience, join a rescue excursion sponsored by your local native plant society.

When mosses are taken from places that won't necessarily be destroyed, the process is considered harvesting. Whether you are retrieving mosses from your own property or someone else's land, take only small sections and leave enough moss in the original locations for the colonies to recover.

Stealing is taking mosses without permission and without regard to ethical standards. Illegal moss harvesting from our protected forests and parks falls into this category. Please do *not* gather mosses from public areas. These recreational spaces are set aside for everybody to enjoy. Preserve our natural treasures so that future generations can appreciate mosses in natural settings. Save mosses; don't steal them!

To acquire mosses for the garden at Silvermont Park in Brevard, North Carolina, Master Gardeners were given permission to rescue from an area slated to become a golf course.

CLOCKWISE FROM TOP LEFT

The shade production section at my mossery is like a fairyland.

Live mosses cultivated at my mossery are available for purchase online in trays or as prevegetated mats.

These tray production samples are on display at the Fairylake Botanical Garden in Shenzhen, Guangdong Province, China.

that grow in smaller mounds as easily as if you were collecting Easter eggs. In other cases, you may have to use a small shovel to dig acrocarpous mosses that have sturdy rhizoids reaching deeper into the ground.

If local mosses are not plentiful in your area or you do not want to go to this much trouble, you can purchase mosses for your garden. Avoid dried mosses sold in plastic bags by huge floral distributors in garden centers or big box stores. Intended for floral or craft purposes rather than landscaping, these mosses may have been subjected to green dyes or even spray-painted to achieve a consistent green color. Regrettably, illegal harvesting to support the floriculture industry is common. If you purchase dried mosses sold in a bag or box called sheet, mood, or cushion moss, you may inadvertently play a role in the domino effect of this complex issue of irresponsible harvesting and corresponding consumer demand.

Most moss distributors function as middlemen between moss harvesters and consumers. Some moss suppliers stage mosses for sale rather than growing them. While some cultivation of mosses occurs in the United States, Germany, and Japan, most dried moss suppliers, even well-known distributors, rely upon obtaining mosses from the wild, not from moss farming. Legitimate and ethical distributors of mosses should avoid questionable harvesting practices, adhere to US Department of Agriculture regulations by obtaining proper licensing and certification, and cultivate live mosses for landscapes.

Your best bet is to purchase live, fresh moss colonies (not dried mosses) from a moss nursery (or mossery) that actually grows mosses for distribution. You can find sources offering extensive selections of common and unique moss species available for purposes of landscaping and gardening. In the United States, certified nurseries are inspected regularly for potential pests or plant diseases that might be of concern to other plant populations. A moss nursery that follows US Department of Agriculture regulations is likely to maintain responsible environmental standards in its harvesting and agricultural practices.

If you acquire mosses from a supplier, make sure to allow enough lead time for thorough planning and for the cultivation of custom orders for living walls or prevegetated moss mats. Do not assume that you can acquire your desired moss species at the last minute. Plan ahead.

To know how much moss to buy, calculate the square footage of your target location. Make sure to acquire mosses that are suitable for the microclimates of your landscape and that you have enough on hand to accomplish your goal. And try to time your moss delivery to coincide with your moss planting schedule. If you get mosses in advance, it is critical to stage them in an outdoor location in the shade and to provide periodic misting to keep them perky. Do not set yourself up for disappointing results or possible failure by stressing mosses before planting.

MOSS PLANTING METHODS

Over the years, moss gardeners have followed simple, basic guidelines in their efforts to encourage mosses in their landscapes. If you merely maintain a debris- and weed-free area, mosses may introduce themselves without further efforts. Chances for success increase if you live in an area where native mosses are nearby because you'll benefit from spore dispersal. As more people have become interested in mosses, more assertive methods for intentional moss planting are appearing on the Internet and in print publications. As moss gardeners venture into new territory in terms of encouraging moss growth, any method that is successful becomes a viable option. While right moss and right place are critical issues, there are a number of ways to achieve desired results.

I have spent years researching and experimenting with a variety of species, planting methods, substrates, soil adjustments (or not), watering regimes, and maintenance routines. While initially following the recommendations of others, I decided to push the envelope and try new methods for myself. If one way worked, it did not necessarily become my only planting choice. I have continued to try a range of techniques, my own as well as others', to determine if one method is better than another. Through this process of systematic yet out-of-the-box thinking, I have uncovered procedures that reliably yield desired results. The researcher in me requires that results can be replicated on a consistent basis under similar conditions. Winning results do not indicate the process is necessarily the only right path or the only method that works. Also, I have learned the hard way that certain methodologies have advantages and disadvantages. For instance, a good moisture-retention substrate or netting to prevent critter destruction could end up being a hassle when it comes to weeding.

As a moss gardening expert, I may make recommendations that differ from those of other moss gardeners. Time will tell which methods really work and consistently yield better long-term results. As far as I am concerned, the field of moss horticulture is in its infancy and all of us still have much to learn. I hope scientists and horticulturists will conduct research that isolates variables such as beneficial nutrient levels for growing particular species of mosses. In the meantime, as more people embrace moss gardening, together we can build the body of moss cultivation knowledge by sharing our successes (and failures) with each other. I heartily support collaborative efforts and exchanges of pertinent information. Since mosses are so stress-tolerant, with adaptive mechanisms for environmental conditions and regional differences, there could very well be more than one "right" way to grow mosses successfully.

My recommendations are based on more than a decade of moss triumphs. The moss planting methods I describe here span the whole spectrum from passively waiting for mosses to grow in to aggressively going for immediate results by laying down prevegetated mats. The method you choose is a matter of your degree of commitment and the time and money you have available to pursue your moss dreams.

My moss mantra

Beyond following the "right moss, right place" philosophy, I urge you to adopt my moss mantra—"water and walk on your mosses." What could be simpler? Do not be afraid to walk on these tiny, seemingly delicate plants. The plant structures are bendable. Walking helps rhizoids of colonies and plant fragments to establish a connection with the soil. Follow this water/walk routine for at least the first month after planting. Once your moss carpet is established, walking on it is a barefootin' bonus!

Letting mosses grow in

Many beginning moss gardeners rely on nearby moss species finding their way into gardens and lawns via spores and vegetative dispersal without any significant human effort. This "let it grow in" attitude assumes that you live in an area where mosses are bountiful and that you have a setting that offers positive microenvironments for moss growth. For people choosing this method, patience is more than a virtue. You need to possess the ability to wait patiently for nature to take its course. Be prepared for a prolonged wait. Folks, we are not talking days or weeks but months or maybe even years to achieve your envisioned results.

Novice moss gardeners, as well as some experienced horticulturists with this optimistic attitude, believe that mosses should "just happen." Even so, some preparation is required and you cannot escape ongoing maintenance chores. When you extend the invitation to mosses to enter your landscape, it is imperative to kick out

competitors early on. You can accomplish the elimination of grasses and weeds in a number of ways. One easy alternative is to cover the entire space with plastic tarps or thick landscape fabric. Blocking sunlight for a few weeks should eradicate unwanted plants. Sometimes I use a weed eater to annihilate vascular plants by cutting them down to the ground. If you have a small space or the grasses or weeds are sparse, I recommend hand weeding. You can grab a hoe to go after stubborn crabgrass or dandelions.

If you find these options too laborious or slow, you can kill rival plants with chemicals or organic substances. Although it seems counterintuitive to use environmentally unfriendly substances with your eco-friendly mosses, many home gardeners and professional landscapers will resort to killing weeds with these substances in lieu of hand weeding large areas. If you are concerned about chemicals being harmful to any existing mosses, do not despair. Usually, a dousing of chemicals minimally affects mosses, if at all. The mosses could experience some temporary stress, but they will most likely rebound. If you feel compelled to deal with weeds in this manner, I suggest you apply the substance in a trial location first.

Once you have conquered the weed competitors and effectively cleared the area of debris, you are ready for mosses to come take up residence. From this point on, keeping up with maintenance activities is essential, so a "let it grow in" attitude does not mean you can be a lazy gardener. You will need to continue addressing debris and weeds for years to come. At some point, if you are lucky, the moss colonies might actually grow so thickly they choke out other plants. In all likelihood, you will deal with a persistent parade of tiny weeds or volunteer plants emerging from the nurturing environment of moist mosses.

A premier example of the "let it grow in" approach is Norie Burnet's moss garden in Richmond, Virginia. She has a reputation as one

This view of the garden from the picture windows of Norie Burnet's charming home took many years to develop.

A garden to restore the soul

Upon meeting Leonora Riddering Burnet (Norie for short), one is immediately impressed with her grace and gentility. It is obvious how magnificent mosses feed her spirit. In her own words, "My moss garden provides a sense of peace and serenity—a oneness with the world. When you are in a moss garden, you will find it restores the soul . . . and souls are around for a long, long time." Nestled in a quiet neighborhood in Richmond, Virginia, her woodland garden features a series of moss walkways radiating from an expansive moss lawn and leading to formal gardens of shade-loving hostas, ferns, and favorite plants like Nippon lily (*Rohdea japonica*).

Mosses hold a sacred place in her heart. In contrast with policies in Japan that discourage any walking on sacred mosses, Norie's policy is to encourage visitors to stroll along her mossy paths on her three-quarters of an acre to find quiet nooks to sit and reflect. Although Norie has been a moss lover and avid gardener since childhood, it was her son Peter who spurred her to let mosses grow in and give up on trying to have a grass lawn in her densely shaded backyard. Besides, as she puts it, "Nothing else wanted to grow in Chesterfield [County] clay." As a schoolteacher and mother of four boys, Norie was just too busy in the 1980s, so she decided to let mosses move in as they pleased, mainly *Thuidium delicatulum* with sections of *Polytrichum commune* and *Leucobryum albidum*. Several decades later, this American moss matriarch has one of the premier moss gardens in our nation.

Norie has learned that waiting for mosses to fill in completely involves some hard work along the way. To this day, she tends her moss garden by herself. Her son Doug puts it this way: "Mom works hard at it. It's not little fairies that come out and take care of all this." Yet, even chores are a pleasure for her. As this spry eighty-something-year-old carefully wriggles out weeds for hours, she finds it "a wonderful way to pull together what is on your mind. It centers you and is almost a spiritual experience."

Over the years, Norie has gained an appreciation for transitions in moss colors. Like many of us, she is drawn to the intensity of green during winter months. She admits that she used to be disappointed with hue deviations. But as an artist, she began to appreciate the seasonal variation of color in her moss garden, and she changed her attitude. Now she smiles as she describes the golden stage of *Thuidium delicatulum* in the spring as looking "like an old camel."

Norie's advice to other moss gardeners is: "Focus and enjoy all aspects of moss gardening. Do not consider it a chore but a real pleasure. If you do not enjoy what you are doing, find another hobby."

RIGHT Mosses and stones make a good combination for paths.

LEFT At Norie Burnet's home in Richmond, Virginia, you can stroll along moss-covered paths of *Thuidium delicatulum*. She encourages people to walk on her mosses.

of America's most successful moss gardeners. This moss matriarch's garden exemplifies how patiently waiting for mosses to grow in on their own can result in moss grandeur. Norie has achieved a magnificent moss lawn as well as meandering moss pathways through her garden. It didn't happen overnight. Photographs that document her progress show significant growth over a period of years. Now, two decades later, her moss mecca is majestic, with major coverage provided by the delicate fernlike moss *Thuidium delicatulum*.

Norie did not sit back on her laurels, though. This octogenarian actively maintains an exquisite garden through hard work and a driving passion for mosses. Burnet notes that debris and leaf removal tasks continue beyond fall months into other seasons. In contrast with some regions where all leaves fall off quickly, she lives in a place where she has to deal with those leaves that hang on into winter months. After high winds or heavy rainstorms, debris removal must happen again. However, for a moss lover like her (and me), experiencing the magic of having your own moss retreat outweighs the drudgery of maintenance responsibilities.

When following this somewhat passive method of moss gardening, expect to wait an extended period before you see any significant results. Be mentally prepared to accept dirt paths and bare spaces until enough mosses take up permanent residence at your home. Burnet put up with muddy paths and waited several years to realize significant growth. Now, decades later, her mosses are sublimely superb. If you consistently keep your mossy areas free of debris and Mother Nature cooperates, it is possible that mega moss rewards will be yours. It just may take a while.

Encouraging moss growth

For slightly more adventuresome moss gardeners, an "encouraging moss" approach will move you a few rungs up the ladder. Your chances of success increase appreciably when you venture beyond just letting mosses grow in naturally and take action to modify environmental conditions to support and promote moss growth. This may include soil modification or use of landscape fabric and from my perspective should certainly include supplemental watering.

Rather than hoping your soil is desirable, why not check it out and know for sure? Take a series of soil samples from various places in your yard. In contrast with samples required for other plants, soil from the surface is all you need. Your primary concern is to determine current soil acidity/alkalinity (pH) in order to know how much soil conditioner you will need to use to raise or lower it. The lower the pH number, the higher the acidity (a pH of 5 is more acidic than a pH of 6). Acid-loving mosses do well if your soil has a pH of 5.5 (although as low as 4 could be acceptable).

The DIY (do-it-yourself) method involves using a simple soil testing kit (available from garden centers). Fill the small vial provided in the kit with water, your soil sample, and the contents of the pH test capsule according to instructions. Shake it up and in a few minutes, you will know your existing pH. Another option in the United States is to send soil samples to your state's cooperative extension service to obtain free or low-cost soil analysis. Soil test reports always include pH status along with other nutrient information.

When you know the pH value of your soil and the acreage or square footage, you can determine how much sulfur or aluminum sulfate to apply to lower your soil pH (if you want to plant acid-loving mosses). You can purchase these items at an agricultural supply store. Alternatively, you can use a packaged soil acidifier for hollies and hydrangeas found at garden centers. Aluminum sulfate is faster acting yet requires a heavier application. Sulfur characteristically is slower to effect a soil balance change, but it lasts longer. In either case, it takes mass quantities of amendments to alter soil pH by

Present pH	Desired pH 6.5	6.0	5.5	5.0	4.5
8.0	1.8	2.4	3.3	4.2	4.8
7.5	1.2	2.1	2.7	3.6	4.2
7.0	0.6	1.2	2.1	3.0	3.6
6.5		0.6	1.5	2.4	2.7
6.0			0.6	1.5	2.1

Pounds of aluminum sulfate per 10 square feet to lower soil pH to desired level

Present pH	Desired pH 6.5	6.0	5.5	5.0	4.5
8.0	0.3	0.4	0.5	0.6	0.7
7.5	0.2	0.3	0.4	0.5	0.6
7.0	0.1	0.2	0.3	0.4	0.5
6.5		0.1	0.2	0.3	0.4
6.0			0.1	0.2	0.3

Pounds of sulfur per 10 square feet to lower soil pH to desired level

even 1 point; refer to the tables here for recommended amounts. I use a handheld broadcast spreader to distribute the pellets before a rain or water them in myself.

With all this said about soil pH, I must confess that elaborate analysis and modification of soil may not be necessary. Mosses can live on all types of substrates in nature and likewise in manipulated gardens.

Beyond soil pH modification, you should not need to supplement the soil with fertilizers. As noted earlier, mosses can grow in nutrient-poor soil, which often is lacking in nitrogen and phosphorus. In reality, the scientific jury is still out on exactly how fertilizers affect mosses in favorable or damaging ways. Michael Fletcher, a researcher in the United Kingdom, has reported that nutrient value or injury could vary by bryophyte species. For instance, he found that one species might thrive with additional magnesium while another might desiccate. Calcium can be particularly detrimental to *Sphagnum* moss. On the other hand, some mosses seem to thrive in greenhouses where fertilizers are consistently applied to other plants.

It is unnecessary to add good topsoil or soil supplements. If you need to add soil to sculpt your hardscape, it is okay to use fill dirt. When transforming my asphalt driveway into a moss retreat, I had a truckload of the "best cruddy dirt" delivered and dumped on top of the blacktop. In the dead of winter, my sons and I planted more than 650 square feet of mosses on top of this inhospitable, unfriendly dirt. After we wore ourselves out shoveling some of this heavy clay-based soil, I resorted to sweet-talking a Bobcat driver to spread the huge pile around. Then I drove my mother's four-wheel-drive car back and forth over the surface to help pack it down before planting. Several years later, the mossy area is thriving—providing a verdant view from my living room window instead of unsightly black pavement.

To increase water retention of an area, some moss gardeners use felt fabric as the planting surface. I have tried a slew of man-made substrates, including felts (thicker commercial grades as well as fabric-store weight), burlap, carpet scraps, sisal rugs, cotton rugs, heavy drapery fabric, recycled carpet fiber, padding used in mattresses, and various types of commercial weed barrier fabrics. I have also experimented with natural substrates, including rocks, gravel, good soil, clay, and grit. After trying all these options, I have determined that a professional landscaping geotextile fabric provides the best man-made substrate option. While permeable, it has water-retention properties. Weeds have a tough time perforating this landscape material, and it does not lose its shape like felt.

For a variety of reasons—weed barrier, erosion control, better moisture retention—you may choose to use a landscape fabric rather than planting directly on soil. If so, you can either add a shallow layer of soil or plant directly on the fabric—in which case adjusting soil pH is a moot point. Saturating or soaking the fabric with an acid-plant fertilizer could be an advantage—but then again, it could be a disadvantage. Even given the benefits of landscape fabrics, my preference is to plant directly on existing soil whenever possible.

Another thing the "encouraging moss" gardener can do is to provide supplemental water. Brief but frequent watering sessions accelerate growth processes and increase your odds of success. Mosses prefer short sessions that occur several times a day rather than drenching soaks. It does not take long—two to five minutes should be more than sufficient. I normally use either the flat, shower, or mist setting on my hose head when watering manually. Frankly, I find the mindless task of watering to be relaxing. I welcome a bit of quiet time to myself. If you can water only once a day, perform this task in the late afternoon after mosses have dried out rather than in the morning when dew is present. From my experience, watering at high noon does not cause burn spots on mosses like with vascular plants. Further, if you want to make like a night owl and water your mosses at midnight, that is okay. In all my years of moss gardening and watering regularly, mold has almost never happened for me—except when I let dead leaves cover mosses too long (more than a month).

Many more watering considerations are discussed in the next chapter. The point for the "encouraging moss" gardener is that a moist environment promotes moss growth, and supplemental watering is a way to help out Mother Nature on this score.

Starting with fragments

Enterprising moss gardeners can encourage growth in new areas by taking advantage of the fact that bryophytes can grow from even the tiniest of leaf fragments. If you want a moss lawn, spreading fragments in your target area boosts the probability of success.

If you want to step up and really encourage mosses, using fragments of species that grow faster or attach quickly is a good bet. *Thuidium delicatulum* is a sideways-growing (pleurocarpous) moss that performs well. Representing the upright-growing (acrocarpous) mosses is *Atrichum undulatum*, which fills in significantly within six months. My favorite moss for fragmentation purposes is *Climacium americanum*, which has sideways rhizoids with an impressive upright growth resembling a little tree. *Climacium* attaches quickly, and the brilliance of its new growth is quite attractive. After breaking up *Leucobryum albidum* or *L. glaucum* colonies into small pieces, you will find out how this moss got its common name of pincushion moss—new growth definitely looks like miniature, rounded pincushions.

Use either your hands or scissors to separate fragments from mother colonies. With

CLOCKWISE FROM TOP LEFT Fragments of *Climacium americanum* are filling in well, and *Marchantia* (liverwort) is aggressively growing without purposeful introduction or encouragement.

My sons, Flint and Carson, helped me spread a huge pile of heavy clay-based soil on my driveway to prepare it for a new life.

Today the former driveway is a magical moss retreat.

pleurocarpous species, you can gently pull apart 1-to-2-inch sections. For teeny-weeny mosses (such as *Ceratodon* or *Bryum*), I ball them up and rub my hands together in a vigorous fashion to separate individual plants from the colony. Certain moss types (such as *Polytrichum* and *Climacium*) have tougher stems, and you will need to use scissors to cut fragments. If any soil is attached to moss rhizoids, I crumble it up, too. I continue sifting through the fragment pile and cutting until I achieve a somewhat consistent size of ½-to-1-inch parts. Note: The more lavishly you spread fragments around, the more you multiply your success ratio. If you skimp, the moss fragments will take longer to grow into plant colonies.

After you have enjoyed your fragmentation frenzy, what is next? You've got it: *water and walk* on your moss fragments. This process will help moss pieces attach to the soil and start growing into plants. Faithfully *water and walk* on new areas for at least the first month. If you *water and walk* on your mosses, they will not blow away easily. Sometimes stretching netting or cheesecloth over fragments is the preferred protocol for helping to hold fragments or colonies in place during the establishment stage. As mosses continue to grow, they will begin forming interconnected colonies. If you perform ongoing maintenance duties (debris, leaf, and weed removal), you should be pleased with your moss fragmentation progress within six months to two years.

Moss milkshakes

The moss milkshake method is a widely promoted but rarely successful method of growing mosses from fragments. This technique involves pulverizing mosses in a blender. First, you need to choose the right moss for your intended slurry; not just any moss will yield the desired results. If you use mosses dried in a bag instead of live mosses, your chances are significantly reduced. Next, I suggest you buy a designated moss blender from a thrift store rather than ruining your own, since mosses can get impossibly tangled up in the blades during the blending process. As for your liquid medium, proponents of the milkshake idea offer plenty of choices, including buttermilk, beer, yogurt, absorba crystals (superabsorbent polymers or SAP), sugar water, and more. Next you slather or paint the mixture on your intended surface, either soil, rock walls, or concrete statuettes. Then prepare yourself to watch it wash away in the first thunderstorm or dry out from scorching sun before mosses ever start to grow.

Having given this method the old college try on more than one occasion and even systematically experimented with it, I have decided it is much too haphazard. I did confirm that if the substrate is more porous, you'll have a better chance. Manufactured stones seem more receptive than natural rocks. But then, my dog liked the taste of buttermilk and licked off the mixture, thus ending my moss milkshake experiment. If all your microclimate conditions are perfect and this mixture survives the challenges of rain, sun, and hungry critters, you could be one of the few who can boast success. While this method has been popularized in the media and thousands of people continue to ask about moss milkshakes at my lectures, I've run into only two people who could verify it worked for them. It sure didn't provide satisfactory results for me.

Planting for immediate results

For gardeners who want to avoid delaying gratification with the let-it-grow, encouraging, or fragmentation approaches, it is time to move up another rung on the moss gardening ladder and intentionally plant expanses of moss. You could complete a contiguous planting project in a matter of hours or, for larger areas, days. By assessing your environmental situation, planning ahead, and following "right moss, right place" guidelines, you will exponentially increase your moss pleasure and reap immediate rewards. That instantaneous rush of adrenaline and self-satisfaction experienced when you plant mosses in a contiguous fashion is absolutely awesome—or should I say moss-some!

If you intend to rework your hardscape, tackle this aspect first. Boulders provide dramatic interest, and you can use them as benches or seating areas. Gigantic rocks can weigh several hundred pounds, so moving boulders is not like rearranging furniture. Before taking delivery of any massive stones, plan where you want them to reside. If you have access to a skidster or earth mover, take advantage of it to remove

LEFT Watering in fragments and then walking on them will help pieces to attach and start growing.

Clear your site and prepare the hardscape for planting mosses. Place heavy boulders first.

topsoil (and weeds). Change the terrain from flat to small hillocks while you have the chance. Obtain smaller-sized stones to use for borders or decorative designs. Some of my favorite rocks include milky quartz crystals and smooth river rocks. Create focal features that draw you into the garden. Plant additional trees, bushes, and flowers before you begin your mossin' efforts. Anticipate where you will locate paths and walkways as well as patio areas and fire rings. If your design includes a water feature, now is the time to get it into position. Install all aspects of your hardscape before you introduce mosses. In other words, get your landscape canvas in order before trying your hand at moss artistry.

For most moss species, you can plant directly on hard-packed, inhospitable soil. You do not need to till or loosen the soil or add fertilizers. Slight undulations in the surface are desirable as niches or channels for moisture accumulation. Slightly moisten the area to observe how water naturally accumulates in spots. Walk around with a heavy heel to create additional minor depressions. Use a rake with rigid tines or a three-pronged hand tool to make slight indentions in the soil as shallow channels. If on a hillside, run grooves horizontally rather than vertically in relation to the slope.

Cookie sheet method

What I call the cookie sheet method yields some pretty good results and doesn't take long. This method is generally much faster than the less systematic fragmentation process. Although you still have a waiting period, you should attain your desired green status in a shorter time frame.

Place hand-sized colonies in a systematic pattern across your prepared landscape, like balls of dough on a cookie sheet. You can use one species or mix up moss types (acrocarps and pleurocarps) to ensure diversity, changing beauty, and long-term survival. The distance between colonies should be one to two times the size of the colony. After you arrange your moss cookies or colonies, green splotches dotting the area will resemble the mass planting of landscaped flower beds. Whenever possible,

To use the cookie sheet method, distribute moss colonies like dollops of dough on a cookie sheet. Fill in surrounding areas with fragments.

plant mosses while they are in the sporophytic state. The wind can be your friend as it disperses spores into new areas.

To amplify the success of the moss cookie method and help sheet colonies grow together faster, scatter fragments and small snatches of mosses between colonies. You can broadcast by hand, taking advantage of a slight breeze to help spread fragments evenly. Another method, reminiscent of fanner baskets used on rice plantations, is to repurpose a daisy tray (a typical plastic carrying tray used in greenhouses). Use a rhythmic motion to sift small moss fragments uniformly through the open partitions along the sides of these lightweight nursery trays. Thicker coverage ramps up the spread of mosses.

For the last stage of cookie sheet planting, water and walk on the moss colonies as well as the fragments. Then, as a temporary measure, it might be a good idea to add a covering of deer netting on top to help hold moss colonies and fragments in place until established. You will find out more about the advantages of netting and some of its drawbacks in the next chapter.

If you perform ongoing maintenance and supplemental watering, you can expect to see some growth within three to six months. Within a year or so, given optimum growing conditions, the moss cookies and fragments will begin to blend together into a cohesive mass of mosses.

Contiguous planting

Contiguous planting is like waving a magic wand unveiling an instant moss garden. Your goal using the contiguous planting method is to end up with solid moss coverage. The idea is to plant colonies right next to each other, either directly on soil or on landscape fabric. If you are trying to achieve a homogenous, seamless expanse, use a single moss species or types that are similar in appearance from a distance. If you want to achieve artistic drama, plant large swaths of the same moss species together with contrasting textures or different greens in adjacent sections.

Arrange your mosses by species along the edges of your intended planting area like dabs of paint on a palette. At a glance, you will be able to assess textures, shapes, and variances in shades of green. Pay attention to any outstanding colonies that have interesting shapes. Colonies sporting colorful sporophytes offer another design dimension to consider as your moss planting progresses. Even with a plan in mind, allow yourself the option of a free-flowing process to achieve a harmonious balance of different moss species, other plants, and garden elements. More than a gardener, you are now a moss artist. If you are an intuitive, right-brained person, you can plant a design with swirls or spirals. If you are more analytical or left-brained, geometric shapes like circles or checkerboard squares can yield major impact. Use a sharp stick or pointy tool to sketch the outline of your design pattern in the ground. If you want to be more sophisticated, you may choose to use landscaping spray paint for a more visible line.

You can start planting anywhere you please. Most often, I choose a primary focal feature such as a tree or boulder as my starting point and work my way out. Locate several moss trays near your target area where you can easily reach them from your perch on your planting pillow (stool, kneeling pad, or whatever). After the first section is in place, turn around to face outward and sit on what you have just planted. Expand your design from this perspective. It is fine for moss gardeners to sit down on the job! Can you think of any other gardeners who sit down right on top of what they have just planted? For mosses, it's a good idea.

From your moss palette along the perimeter, choose your first colonies. Pick up the first section in one hand; vigorously rub your other hand across the green side to fluff it up. Hold it over the planting area so that any fragments will land in desirable positions. Turn the colony upside down and give it a good shake as well. If there are any mud clods or ugly debris, remove from the top of the colony. To achieve a pristine and clean appearance, delicately brush tiny moss species with your designated moss toothbrush.

As you begin planting, interleaf the edges of pleurocarpous moss colonies that grow in horizontal mats, using an over-and-under method along the perimeter. After interleafing, press along the seams of these opposing sides to help them interconnect. The irregular shapes of natural moss colonies may present some challenges as you begin to fit the pieces together. However, as you start to connect with the mosses and get comfortable with the planting process, you will know which one to use next. Colonies start fitting together like a puzzle or patchwork quilt. If you acquire mosses from a nursery, the colonies may have assumed the shape of the tray in which they were cultivated. You may want to work with a more ragged edge. If so, tug gently to stretch colonies and pull out into ruffled edges. For geometric designs, the distinct edges of tray-cultivated mosses may be advantageous. You can also cut distinct shapes with scissors. Tuck mosses around feature accent plants or make a hole in the moss mat for them. Continue this interleafing process with

FROM TOP TO BOTTOM

Rather than the homogenous effect of one species, I prefer the varying textures of several species growing together in harmony.

It's okay to sit down on this job. Look at the spectacular results achieved from planting mosses in the cracks of this rock wall.

Although not a necessity, small twigs can be used to hold overlapped, interleafed edges in place during the establishment phase or for repairs.

your pleurocarpous mosses until you are ready to switch to another moss type. Using small twigs can help hold edges in place.

For acrocarpous mosses, push colonies close enough to touch and nestle into each other. Consider using the layering effect of taller mosses in the background and shorter species in the front. Press colonies firmly into place. As my daddy taught me, "Do it with authority." You do not need to be gentle with mosses. Although it might seem mean, sometimes I punch down into and stomp on huge *Polytrichum* colonies to help them securely attach.

Punching massive colonies of mosses into the ground is way-out-of-the-box thinking when it comes to gardening. But I have found other unconventional methods that work with mosses as well. Since I've always loved rolling around on the ground, it is not unusual for me to lie down on soft moss carpets and roll around—yet another way to encourage effective rhizoid connections. Flexible moss plants and dainty reproductive sporophytes bounce back to an erect position. Of course, finally, I recommend conventional walking on your new moss areas. Landscapers may choose to use tampers or heavy rollers.

As you move toward your ultimate goal, pause between stages to admire your artistic results. Your sense of accomplishment will fuel your continued moss gardening efforts. Excitement will build as you anticipate the final results. As the moss transformation occurs before your eyes, you will fully understand why the contiguous planting method provides such immense and immediate gratification. If you try this technique, you will find it absolutely amazing to orchestrate your own instant moss makeover. To go from a blank canvas to a finished moss masterpiece in a matter of hours or a few days is certainly a marvelous moss experience.

Planting mounds in a contiguous fashion is simple. Place colonies in close proximity to each other without any gaps.

Rolling out the green carpet

While strutting across the red carpet is the dream of Hollywood stars, moss gardeners envision strolling along their own green carpet. We may be willing to wait years for this ultimate event. We may choose to speed up the process by encouraging moss growth and taking proactive measures. But wouldn't it be wonderful to roll out your moss carpet like grass lovers roll out sod? For instantaneous gratification, what could be better?

At my mossery, I have been producing prevegetated moss mats with standard dimensions of 6 by 6 feet (36 square feet) as well as custom shapes and sizes composed of either shade or sun species. Splashing design concepts and geometric patterns add extraordinary drama to this easiest of all methods for achieving moss reality in your garden or lawn.

It is simple to roll out your green carpet. You do not even need to remove weeds from your intended area—just clear debris and unfurl your mat. A series of mats can be interleafed for larger-scale projects. Using scissors, you can easily cut prevegetated moss mats into irregular or geometric shapes for garden features or long strips for paths. Secure mats in place by strategically placing landscape pins (or garden staples) along the edges. Finally, remember to water and make sure to relish your own royal stroll on your new emerald carpet.

TOP You can use hand scissors or a rotary scissor tool to cut prevegetated mats to fit the space or to create shapes, even words for walls.

Easy installation and instantaneous gratification is possible with prevegetated mats.

LUXURIOUS LAWNS, PREMIER PATIOS, AND RAVISHING ROCK WALLS

Yearning for a luscious moss lawn and softening the hardscape of a stone patio or rock wall are among the most popular mossy dreams. All are within reach—it is just a matter of time and your degree of commitment.

Getting started on a moss lawn

If you have been fighting a battle against mosses in grassy areas, you may be ready to give in and let the mosses win. The first step is to stop planting grass seed and adding lime to your soil. The next critical step in the process is to remove all debris down to bare soil. Lest you end up frustrated, do not bite off more than you can chew. Grandiose schemes of large moss expanses can be expensive and/or time-consuming to plant and keep maintained. To replace acres of grass may prove to be a monumental task. Be realistic and scale your moss lawn endeavors to suit your pocketbook and your time availability.

Rather than trying to mimic the homogenous look of grass, consider introducing a variety of moss species for a range of textures and shades of green that provide dynamic and diverse appeal throughout all seasons. While you can avoid the use of fertilizers, herbicides, and pesticides needed for a lush grass lawn, you may need to provide supplemental watering. Frequent but brief watering sessions suit moss lawns, so you can at least avoid long drenching soaks and thereby reduce your water consumption.

Adding moss to horizontal hardscape surfaces

The pièce de résistance or crowning touch to an elegant patio can be achieved by adding mosses in the spaces between the stones, bricks, or pavers. This extra mossy embellishment adds a green softness to the harshness of hardscape surfaces. Beyond patio possibilities, you can use mosses in permeable driveways and walkways. Pebble stone driveways made with epoxy resins are great candidates for moss treatments. With mosses threaded through the gaps, your stones will pop as impressive elements. Taking this final moss step using fragmentation or contiguous planting methods ensures the wow factor of your outdoor living areas.

Choose moss species that want to grow in cracks anyway. As mentioned earlier, several species found worldwide—including *Bryum*, *Ceratodon*, and *Entodon* species—tolerate the high heat of reflective surfaces. The first two species really prefer direct sun locations while *Entodon* thrives in either sun or shade. In my experience, *Bryum argenteum* will dissipate away if located in a walkway that is too shady, too soggy, or just not hot enough.

If your impervious surface is located in the shade, you have many species options. Many people may choose acrocarpous mosses (upright growers) since they do not have the tendency of pleurocarpous mosses (sideways growers) to creep onto pavers. Keep in mind that the more porous your hard surfaces, the more likely it is that mosses will start to grow on them. Slate and bluestone resist moss growth (although moss could possibly attach to these over time), while in time limestone or granite rocks display expansive growth.

Fragmentation is by far the easiest method for filling in the cracks. If you want more immediate transformation, use the contiguous method. Be forewarned, planting small colonies in vacant spaces between stones can be tedious. It takes more time than you might think when you are placing so many mini-colonies. Various substrates are useful for leveling and setting the base under stones or pavers. Stone dust inhibits drainage—a plus for many moisture-loving mosses. What is called crusher run, with varying sizes of crushed stone, is more granular than powdery and allows more drainage. *Leucobryum* species like good drainage. If you

LEFT To plant the cracks of a wall along a pond's edge, I donned waders and climbed right in.

The effect of combining mosses with flagstone for a patio or path is quite desirable.

Living moss walls

Living walls are gaining in popularity as urban dwellers yearn to add green to their concrete jungle. Frequently, *Sphagnum* moss (peat) is used as a growing medium in pockets or bags designated for other plants. Of course, my idea is to feature the beauty of mosses on green walls. To that end, I've developed a process for planting pleurocarpous mosses on lightweight, portable panels. After cultivating the moss panels in a horizontal position, I frame them up and hang them in a vertical position—swaying in the wind or attached permanently to a wall.

As with all my experimentation, the true measure of success is how well panels withstand the test of time. One night during a particularly intense thunderstorm, I heard loud knocking noises and worried about my experimental moss panels banging around. The next day, I discovered that one corner had slipped off its hook, causing the flapping racket. To my delight, not one bit of moss had fallen off! I'm pleased to report that Mountain Moss's living moss panels have proven to endure high winds, heavy rains, and hailstorms.

LEFT Once moss rhizoids have attached to the substrate, prevegetated wall panels or moss mats can be hung vertically.

RIGHT Moss wall panels are a great way to green an urban balcony or a rooftop in the city.

are dealing with better-draining grit or sand, you may want to add a mud pie mixture of clay, pine needles, and leaf compost that will suit other moss types. If you are using live mosses (acrocarps in particular), you will benefit from the soil already attached to rhizoids.

Having to deal with any type of stone grit or sand is downright messy. Keep your whisk broom handy to maintain a clean work area and minimize peripheral grit drift onto the top of newly planted mosses. Before you begin, spray rocks clean of all grit. As you work, sweep up extraneous grit. Using a trowel, dig a shallow trench for planting mosses between stones. Firmly place mosses with authority. Sweep again using a hand broom and dustpan. Water and walk on your moss and stone patio, path, or driveway area to conclude the process. Your hard labor will pay off. Adding mosses offers an intangible and timeless quality to any outdoor hardscape.

Softening rock walls

Hardscape elements certainly look better after moss enhancements. Consider adding species that like to grow on rocks or bricks to your walls. Choose mosses that tolerate the heat emitted and retained by the substrate. Planting is quite simple. Take small colonies or fragmented sections and nudge into the mortar cracks or natural niches. Keep a bit of colony soil attached or make up a supplemental mud pie mixture. Water down the rocks first; then poke mosses and soil mix firmly into place. Water again and push again. Another option is to glue mosses in place. Since rhizoids don't provide nutrient transport to the mosses, you can use masonry glue or a two-part epoxy.

your thriving moss garden

MAINTENANCE AND TROUBLESHOOTING

Keeping your mosses healthy and happy is essential in achieving lasting splendor. I wish I could say that mosses require zero maintenance, but the reality is that magnificent moss gardens necessitate ongoing attention.

You can expect a number of requisite chores as you maintain the mystical appeal of your moss kingdom. The routine is relatively easy albeit a bit labor-intensive at times. Supplemental watering is highly recommended. Any litter or debris must be picked up. Repairing damaged areas is crucial, too. Getting rid of weeds is another continual concern.

There is no way to get around putting forth some effort to sustain and encourage positive moss growth. However, your reward of thriving and healthy moss colonies will be worth the effort you exert in maintenance activities. My best advice is to stay on top of essential duties lest the undertaking become overwhelming. Most of all, I hope you will find as much pleasure in the meditative aspects of maintenance as I do and enjoy the process as your own road to enlightenment.

UP CLOSE AND PERSONAL WITH YOUR MOSSES

Paying attention to maintenance needs and following through with required actions is the basis of good gardening. It starts with being observant. Walk through your moss garden and take note of obvious needs. Make a to-do list of areas of concern and determine your clean-up plan. It is not necessary to write it down—just make mental notes. From a panoramic perspective, you can easily see fallen leaves and broken branches resulting from high winds or heavy thunderstorms.

In assessing conditions, you may see smaller debris such as acorns as well as pockets of accumulated conifer needles. If you are up close and personal with your mosses, you will discover weeds at their earliest stages of growth when removal is easy. It does not take long to develop a keen awareness of how environmental conditions impact mosses, particularly the clear visual clues of wet versus dry. As you refine your observation skills, you will know from a glance when your moss garden needs watering just from the shade of green exhibited.

Walking on your mosses

You may want to tiptoe through your tulips, but as we have already established, walking on mosses can be strategically advantageous. You need not walk gingerly in your moss garden, although you do not want to wear shoes with treads or clodhopper boots that might scuff up or dislodge moss colonies. It is okay to walk around on your mosses anytime, but it is especially valuable after watering, fragmentation planting, and repair work. The point of moss walking is to firmly press moss plants and pieces into damp soil as a means of encouraging expansion into new areas and/or "overseeding"

PAGE 190 *Dicranum scoparium* colonies may fall victim to damage by squirrels or birds. Netting may be your solution.

OPPOSITE Keeping a moss landscape healthy requires some labor but is well worth the effort.

RIGHT Not just a concern in autumn, leaves, pine needles, and other litter may accumulate in other months, too, and should be removed on a regular basis.

Regular policing for debris and weeds provides early intervention before problems get out of hand.

places where you have used the fragmentation planting method. Fragments are less likely to blow away if embedded in soil.

When repositioning moss colonies to fix up areas visited by critters or damaged by the effects of weather, walk on repaired patches in a steady manner. Avoid any fast, twisting foot motions. Make sure to step carefully along the edges of pleurocarpous (sideways-growing) moss colonies during the repair process. Personally, I take repeated baby steps across the entire area. Water colonies and walk on them one more time. As you should know by now, moss walking helps rhizoids attach to soil more effectively and faster.

While walking can be a good thing, too much heavy foot traffic (from visitors, for instance) might result in some wear and tear. Periodically, you may need to address this maintenance issue by adding more mosses to barren spaces in favorite paths. If you want defined moss paths that can withstand regular walking by a steady stream of garden guests, try using pebbles or small gravel as the base. I have discovered it is necessary to place an underlay of landscape fabric to keep small stones from eventually disappearing into the soil. You can toss moss fragments among the pebbles and walk with confidence that moss growth in the path will happen eventually with little or no negative impact from foot traffic.

Visible clues

As you stroll around your moss garden, noticing visible changes in appearance will help build your understanding of how best to care for your mosses and maintain a pristine appearance. Knowing whether the difference is a reaction to environmental stresses or just a normal transition will guide the need for remediation. Shriveled moss plants usually are indicative of a dry state. Your mosses are probably just in need of a drink. A squirt of water will change them back to a more pleasing look as rehydration occurs quickly. If mosses look normal again after watering, you can rest easy. If moss colonies do not rebound to a lush state, there might be cause for concern, but not always. As

Over time, gravel paths in your garden will start to green with mosses that self-introduce from nearby colonies. Speed up this process by tossing moss fragments into the path during weeding.

you may recall, color changes can occur during normal transitions and reproductive stages.

Of course, you need to be aware of what is normal for each of your moss species in order to differentiate variances. Photographs are valuable references in the learning process but hands-on gardening will help you develop skills for detecting both major and subtle deviations. To establish a point of reference, you should examine features when mosses are in their ideal state of wetness. Leaf characteristics and color differences can serve as points of reference in determining whether your mosses are happy campers or not.

Color change as an indicator of health

Green glory is the goal of many moss gardeners, but knowledgeable mossers realize that a full spectrum of greens—from deep, dark forest green to exotic emerald to lime with golden overtones—as well as bronzes and other brilliant colors exist in the world of bryophytes. Various moss species have dramatically different greens from each other, and even a single species may exhibit an array of green hues in response to climatic conditions and life cycle stages. After a good rain or during new growth spurts, the green will be more vibrant. In comparison, dormant, old, or dry mosses seem to be duller in color.

Less appealing colors may indicate desiccation or imminent death, but then again, they may not. The burnt orange tops of *Atrichum* species are the male cups, not sick plants. The melon tint of *Hedwigia* species with distinctively darker leaf stems does not mean they are ill. I am not particularly fond of brown but I have learned to accept it as a normal color for certain species, including *Polytrichum*, *Atrichum*, and *Climacium*. In mature *Polytrichum* colonies, the older growth is a rusty brown. Green shoots originate from the top portion of this older brown during the next growth spurt; each year I notice at least an inch of new growth.

On the other hand, certain mosses do not like wet feet, and when *Hypnum* stays too soggy for extended periods, its yucky brown state proclaims its distress. When silvery *Bryum* turns a dull gray, it is most likely suffering from heat and sun exposure. Extremely tolerant of these climate extremes, this minute moss species most often recovers when wet and humid conditions reoccur. Regrettably, when *Leucobryum* plants turn gray, the colony is most likely heading toward death. Remarkably, when a *Leucobryum* colony turns black, I've observed spiky new green growth within a month, a dramatic recovery.

When *Hypnum* and *Thuidium* mosses turn yellow, they are not sick but are reacting to more sun exposure. I have had yellow *Hypnum* mosses turn back to solid green just a few days after relocating them from sun to shade! While mosses that turn a bright yellow may still be healthy, a sallow, drab yellow is evidence of an unhealthy state. These suffering mosses may take a long time to recover—perhaps as much as six months to a year, maybe never. When *Dicranum* moss turns yellow from the center of the colony, it is on its way to the grave. Once when I temporarily staged members of this species in a hot, sunny location, lush green colonies baked like buns in the oven. Within a few hours, they turned a sick, dull yellow color. Another time, I watched with dismay as *Hypnum* moss became devoid of all color, turning almost white after only one day in direct sun. Even after being moved to the shade, these frazzled colonies took months to return to a green state.

Changes in color may also be a result of dormancy, when growth slows down. After their sporophytic stage, *Ceratodon* plants turn a lackluster shade of brick red—almost brown. It looks like the colony is surely dying. Just wait a few weeks, though, and you will be rewarded with luxuriant growth once again. But if moss colonies are not going through a dormant stage, brown, gray, black, white, or dead yellow colors

Large *Polytrichum* hummocks may display long brown tendrils of older growth with new green tips. The brown is normal and does not necessarily indicate a problem.

can indicate significant stress or the potential of death.

Before tossing in the towel, give these impacted mosses a chance. One alternative is to change locations, transplanting mosses to a shadier or sunnier spot. Perhaps you need to modify your watering regime by eliminating, reducing, or increasing watering. If you have resisted supplemental watering, you may need to give in and water your mosses.

In summary, be observant of your surroundings and aware of forces that impact your mosses. Watch for color change as a key indicator of the need to modify situations. Determine the best method for resolving the particular concern. Wait a while (several weeks or months) and see if revitalization occurs. Don't give up.

WATERING CONSIDERATIONS

Surviving or thriving—which do you choose? If you want consistently lush mosses, you need to provide supplemental watering. Most gardeners recognize watering as an essential chore for flowers, plants, and trees to survive. Vegetable gardeners dutifully give daily drinks to their edibles. Irrigation is commonplace in farming. It just makes sense to water, doesn't it? Without consistent humidity and natural rainfall or the intervention of gardeners who intentionally provide supplemental watering, mosses are subject to drying out. Fortunately, parched and thirsty expanses will rebound quickly when Mother Nature douses thirsty leaves with rain or you intercede with a dose of rejuvenating water.

If you live where humidity is low and temperatures are hot, supplemental watering is essential. Mosses, and even vascular plants, may require supplemental watering or misting if daytime and nighttime temperatures vary only slightly (hot at night without any cooling down). Even if you live where the annual rainfall is high, dry days or periods of drought still occur. While Transylvania County, North Carolina, has an annual rainfall of up to 105 inches, I still follow a supplemental watering regime. In colder months I relax my routine, but I do still water my mosses on warm winter days. I leave my hoses in place during freezing weather but detach them from the outdoor spigot; I reattach them as needed. I was surprised to discover that wet snow is not as wet as water. When snow melts, mosses can be rather dry, believe it or not.

If you are concerned about water conservation, mosses do not require nearly the amounts necessary for other plants. A short session will rehydrate mosses quickly. It does not take much more than a sip for mosses to be happy campers. Brief sessions should last between one to five minutes. In warmer summer months, several sessions each day may be valuable. Misting irrigation systems provide more intensive overall coverage and faster saturation. Moss watering times probably will be shorter with this technique than with other supplemental watering methods.

Moisture is key to encouraging prosperous mosses. Knowing when to water and for how long will be central to your success. Before we jump to the how-to, though, a discussion of water quality and how it may impact your moss gardening methods is warranted. For that matter, let us examine properties of rain and snow as well. Understanding the interconnectedness of environmental factors will help you become a better moss gardener.

Water quality issues

Water quality can be a legitimate concern to moss gardeners in some regions or cities. While I personally have never experienced any ill effects from using municipal water sources to irrigate my garden, other moss gardeners report issues such as the accumulation of salts on moss plants. Hard water contains magnesium and calcium, which form salt deposits. Even if you have a well, mineral content can have a negative impact. Contact your local Soil and Water Conservation District office to find out about water concerns related to agriculture and residential landscaping in your region.

In America, urban water systems must meet Environmental Protection Agency (EPA) standards for drinking water quality. Various states have implemented regulations with higher standards than required by the EPA. So, how clean is your water source? Have contaminants compromised water quality? If you live in a metropolitan area, you can find out about your own water quality by checking out the EPA's consumer confidence reports.

Do chemicals such as chlorine and chloramine, used in processing tap water for human consumption, affect mosses? In my neighborhood, 450 homes receive water from a spring-fed well that meets EPA standards for a

Snow doesn't harm mosses, but it doesn't necessarily irrigate them, since it does not contain as much moisture as water.

community water supply. My tap water has the typical amounts of chlorine used to disinfect water. I am not too concerned about such trace amounts, particularly since I have watered for years without damage. If you poured full-strength chlorine bleach directly on any moss, it would definitely die. Dumping tap water from a watering can, bucket, or hose without diffusing it could cause chlorine damage. Misting is a better idea because it gives the water lots of surface area from which to lose the chlorine gas before the water reaches the moss.

If your water quality is questionable, do not despair. There is a sensible solution. You can add a water purifying unit to your home tap water source. A water filter can remove chlorine, iron, mercury, lead, and other undesirable volatile organic compounds before tap water reaches your plants. As the use of graywater (water collected after use from baths and sinks) gains popularity as a method for meeting landscaping needs, be aware that salt and soap residues may harm mosses, or they may not. The impact of graywater on mosses has not been adequately tested. Recycling graywater through a filter before it hits the garden resolves any possible issues. If you are apprehensive about water quality issues, alleviate your fears and prevent problems by investing in a water filtration device. Problem solved. Rarely is distilled bottled water needed.

Rain and snow are not exactly pure, either. Contaminants in the air can get trapped in raindrops and snowflakes. Rainfall leans toward the acidic side, with a pH of slightly less than 6. The Extension Service at Clemson University explains, “High rainfall leaches basic nutrients such as calcium and magnesium from the soil, which are replaced by acidic elements such as aluminum and iron. For this reason, soils formed under high rainfall conditions are more acidic than those formed under arid (dry) conditions.” The pH of tap water, well water, rain, and snow affects soil pH, which in turn can affect your mosses.

Acid rain is the result of natural sources (volcanoes and decaying vegetation) and emissions from fossil fuel combustion (vehicles and power plants). Acid rain has a pH of 5 to 5.5, ranging as low as 4 in heavily populated areas with lots of cars and industrial sites. While acid rain formation occurs in cities, the wind carries it hundreds of miles away into the countryside. If limestone is prevalent, the soil balances out the acid rain. However, where limestone does not occur, soil acidity levels increase. I live in mountains where crystal waters flow, but decades ago acid rain originating from northeastern cities started to impact vegetation in North Carolina. Air and water pollution are concerns in these hills. Acid rain is considered harmful to our environment, but in the case of mosses that prefer acidic conditions, observation indicates that these resilient plants might actually benefit from increased acidity via rain or snow.

Collecting rainwater for irrigation

Rainwater harvesting is a great sustainable option for providing water to your mosses. Check regulations in your state; you may be appalled to find out that rainwater harvesting is restricted in your state, particularly in western states, or you may be pleased to find that a tax credit is available.

Depending on where you live, you may collect acid rain—but if higher acidity is a good thing for mosses, the pH may be perfectly fine. Realize that as rainwater rolls over your roof, it accumulates not only dust and soot from the air but also chemicals from roofing materials. Mosses growing on roofs are commonly sighted in my region, so I believe there is not much need to fret—in most cases. However, be alert to harmful effects of watering with rainwater that has channeled through copper gutters and downspouts.

Assuming you prefer using a water barrel or larger cistern, the biggest issue is to facilitate distribution rather than depending on that heavy watering can. Carrying heavy water buckets is for the birds—or mules. If you truly want to go off the grid, use solar panels to power a pump. If electricity is accessible, you can use a small pump to power a sophisticated irrigation system or a simple garden hose and water sprinkler combination. You may need to add one or more filters at the water collection source to remove sediment and debris. Keep any holes on the top covered with a fine mesh so that inquisitive or thirsty critters do not drown in your barrel. Take precautions to prevent mosquito infestations in standing water.

While rainwater collection is a practical solution, do not forget the aesthetics of your garden. Plastic containers are available in a variety of colors and textures. You can clad your barrel or cistern in a natural material to reduce its obtrusiveness. Consider using bamboo fencing or twig blinds as camouflage.

When and how to water

I have made it clear that I believe in the benefits of supplemental watering. Mosses hydrate and dry out more rapidly than other plants. Do not assume that a heavy rainfall one day eliminates the need for supplemental watering the next day. Relying on rainfall alone may yield inconsistent results and compromised beauty. If you want to encourage faster growth and maintain an intoxicating appearance, you can help out Mother Nature by serving up daily drinks. The key is to provide the right amount of water at the right time.

If you live where dew accumulates each morning or mists are common at night, watering in the afternoon is ideal. If you live in a location that has consistently low humidity (dry air), watering at any time of the day or night is advantageous. Actually, if your mosses are dry, it is okay to water at high noon or even midnight. Mosses do not care. I have watered when the midday sun was most intense without experiencing any spotting or leaf burn. And, despite other gardeners' warnings about mold, I have never had to deal with any mold issues as a result of my moonlit watering sessions. By the way, moonlight aids the growth of mosses; like other vascular plants, mosses have indole-3-acetic acid, a naturally occurring plant hormone that stimulates growth at night. This scientific fact confirms stories I have heard from old-time moss harvesters about beneficial lunar influence.

Moss gardens located in geographic regions with excessively hot temperatures, low annual rainfall, and blustery winds will decidedly benefit from supplemental watering on a regular and consistent basis. Even in locations where all the conditions seem perfect, there could be times when they are far from ideal. When you

have multiple days in a row without rain and/or temperatures exceeding 85°F, be especially conscious about watering needs.

The easiest and least expensive way to keep your mosses happy is to use a garden hose and multi-array spray nozzle. If you have the time, this method of hand watering is personally rewarding because you really connect with your mosses as you move around the garden. It is bewitching to watch the almost-instant transformation as mosses become saturated.

I realize that some people just do not have the time to meet daily watering needs. The reality is that I also am too busy sometimes. To alleviate this issue, I urge using a sprinkler on a battery timer system so that hands-free watering happens on a regular basis. Instead of using a timer that has fixed six-hour increments, I recommend a model that allows you to program sessions for different lengths of time, specific times of day, and various zones. For instance, I have three oscillating sprinklers placed in my moss garden with watering sessions scheduled at 10:30 a.m., 2:30 p.m., and 5:00 p.m. Based on

Overhead watering (several two-to-five-minute sessions each day) leads to faster growth, better reproduction, and healthier moss plants.

the varying sun exposures of moss locations, the length of the watering sessions differs. In the shadiest area, the sprinkler is set up to run two, maybe three minutes. In direct sun, sessions last for a maximum of five minutes.

Use your judgment to determine what time to start sessions and how long they should run. Literally, one minute may be sufficient. Usually run times of two to four minutes are appropriate to achieve a fully saturated state. If mosses are dry to the touch between sessions, decrease the amount of time between sessions. If mosses are soggy, increase the time period between sessions and/or reduce the number of minutes of watering. Occasionally, you may want to relocate the placement of your sprinkler heads to achieve better coverage. As a jumping-off place, try using the same start times I use. If it stays hot until sunset, another session might be in order. Just like a kid who wants that final glass of water before bedtime, mosses like a sip as dark envelopes your garden. Also, as seasons change, you may need to adjust start times and run times. As expected, more watering may be necessary in summer while less might be needed in winter when evaporation slows down in colder weather.

When it is raining and extra watering is obviously unnecessary, you can manually turn off the timer for twenty-four hours using the rain delay button. The system will return to its routine the next day. However, if you think your mosses are still too wet, reduce the frequency and amount of water provided. Delay starting the system back up for another twenty-four hours if your mosses are oversaturated. Many garden moss species do not want to remain in a waterlogged state; in fact, some drying out can be a good thing, allowing cells membranes to repair and rejuvenate.

Even so, as a general rule, erring on the side of too much watering is better than letting mosses sit dry for days or weeks. Remember, mosses need moisture for certain aspects of reproduction. Those little sperm need water for swimming to the egg for fertilization. Gemmae balls cannot splash out if the cup is dry. Besides, to me, mosses look much better when hydrated. The point is to find a balance. Know your mosses and recognize the signs of stress or desiccation. And be forewarned that consistent overwatering can be bad news. Still, if you turn on a sprinkler and absentmindedly let it run for hours, do not get too upset. This "oops" mistake will not hurt mosses. Just try to follow your basic troubleshooting guidelines: Water mosses if dry. Do not water if soggy to the touch.

In colder winter months, when pipes are subject to freezing, you may have no choice but to suspend watering and disconnect your irrigation system. However, if an exceptionally warm day happens in winter, I might water. It may not be a critical need for the mosses but it makes me happy. Even in the wintertime, you can enjoy the magical transformation of green color change as saturation occurs (if your mosses are not covered in 3 feet of snow).

If an installed irrigation system is an option for you, advise your professional irrigation specialists of the differing needs of mosses versus other plants. Inform them that mosses prefer misting irrigation emitters (spray heads) instead of underground soakers. Let them know that mosses favor frequent, short mists rather than a long, drenching soak. Make sure they assign separate zones for moss sections with the capability of programming specific times of the day and different session lengths as needed. Beyond this bit of moss advice, let them handle installation and calculating proper water pressure. Licensed irrigation contractors are familiar with the complexities of considering water flow rate, irrigated area, and system efficiency.

As an ongoing troubleshooting measure, you should monitor the irrigation system. After installation, run through the scheduled cycles to make sure the timing is right. The next watering session should happen when mosses are heading toward the dry state, not still moist.

TOP Installed irrigation systems or a garden hose with sprinkler on a timer relieve gardeners from personally performing this daily task.

If you have small areas of moss, watering them yourself provides a relaxing and pleasurable experience. Use mist, flat, or shower settings on the spray nozzle.

Before the crew departs, get some hands-on instruction on using manual override functions, particularly the rain delay button. Subsequently, as part of your troubleshooting routine, you should pay particular attention to the need for adjusting timing and reporting any system failures. Whenever necessary, promptly alert your irrigation contractor that a service call is required.

Whatever watering method you decide to use, walking on your mosses afterward is advantageous. Watering and walking are good maintenance habits for successful moss gardeners.

LITTER REMOVAL AND REPAIRS

You will find that a majority of your maintenance chores in a moss garden are a result of other plants—not the mosses themselves. Your maintenance efforts will be labor-intensive with leaves in autumn. But the need for litter removal continues through other seasons as vascular plants go through their life cycles, eventually dropping flowers, seeds, nuts, and fruits. Not surprisingly, at any time during the year, you can expect to conduct pick-up patrols after a thunderstorm associated with heavy rains and high winds. Dale Sievert in Wisconsin refers to this process as "policing the area." As you move through different sections, pick up debris as you keep an eye out for other troublesome spots such as dislodged colonies.

Next, using a leaf blower is helpful as you move forward with a comprehensive housekeeping plan for your moss garden. While it is tempting to replace dislodged colonies immediately upon discovery, wait until after your initial power blowing session. Otherwise, you could have an "oops" experience and end up doing double duty. In an effort to reduce tasks related to debris and critter damage, you may want to consider using netting as a preventive measure and maintenance solution. Leaf control can include a combination of blowing and netting methods. But before you net, you need to tend to the business of repairs.

While raking is a traditional method of removing litter and leaves, moss gardeners should shy away from this garden tool. The tines can easily grab and dislodge mosses. If you are determined to use a rake in your moss garden, you will need to develop a light-touch technique.

Collect your leaf and pine needle litter to begin your own dedicated compost for moss gardening. It might come in handy when you need some rich humus or want to make a mud pie mixture. Do not mix in vegetables or other yard debris, particularly your discarded weeds. Leave those materials for your other garden compost piles.

Leaf blowing

I am a firm believer that established mosses should be able to withstand power blowing even on the highest setting. After years of due diligence, I have found that blowing when leaves are wet can be quite advantageous. Leaves and smaller debris effectively blow away but most established moss colonies hold their position. If you power blow when the area is dry, you may be horrified to see mosses swirling in the wind with the leaves. I recommend using an electric or battery-operated leaf blower so as not to contribute to air pollution.

For removing litter while watering with a garden hose, I use the power full-force stream setting on the spray nozzle to "water blow" leaves to perimeter areas. Rapid, jerky movements work best when in water-blow mode. This intense water spurt works well for reaching the other side of the garden. Do not use full force without keeping the water spray moving all the time, or you could power squirt your mosses loose.

After blowing, your moss garden will look much better—at least from a distance. Now, do another walkabout focusing on removal of small accumulations of embedded twigs, acorns, and the like that did not initially blow away. Armed with a tall broom or handheld whisk broom, sweep your mosses (yes, I really did say "sweep"!), and then hand remove the last remnants of litter as needed. True fanatics can bring along a toothbrush for fine-toothed cleaning of the tiniest moss species.

Getting rid of needles from conifers can be more challenging than dealing with leaves. A combination of power blowing, sweeping, and hand removal methods is necessary. Try swishing your hand back and forth to loosen needles and then pluck them out. If you are a neat freak, you can spend hours obsessing over pine needles. Like me, you may need to accept that it is a nearly impossible task to remove every single needle; your mosses can handle a few strays.

It is unnecessary to remove leaf or needle litter the moment it hits the ground. Dust particles from leaves and conifer needles provide some nutritional value to mosses. In addition, leaf litter can protect mosses from sunburn

Vexing vascular plant chores

In the temperate climate where I live, the attractiveness of flowers and ferns is short-lived in comparison with mosses. In the fall, I am busy pulling off seedpods, pruning away unsightly dead stems, and removing dried-up foliage and fern fronds. With my abhorrence for ugly, brown, leftover plants, I grouchily remove remnants of dead and dormant vascular plants. My banter of irritable comments is interspersed with accolades of praise for the superiority of my mosses. After this seasonal clean-up is completed, I am proud to showcase my moss garden. Uninterrupted green during the winter months is quite divine.

Expect to deal with litter chores throughout all seasons.

TOP Power blowers (preferably electric) can be used to clear leaves off mosses. Blow when wet, not dry, for best success.

Like moss gardeners in Japan, I use a handmade broom to sweep away debris.

and drying out. Sometimes when you blow leaves away, the mosses trapped underneath are actually greener than mosses not covered by leaves. It is just fine to let leaves remain for short periods. However, if you neglect piles of leaves for more than a season, you are inviting mold or moss death. A thick covering of leaves prevents photosynthesis and is a sure way to kill mosses eventually.

Trees, bushes, and flowers are "litter bugs" in other ways. Let us use oak trees as an example. Some deciduous species do not drop their leaves until spring. By late spring, dead flowers (catkins) that are brown and stringy drop off, creating brown blankets over your green mosses. In fall, especially during a mast year, acorns (fruit) can bombard at alarming rates so that thousands may cover your mosses. You had better blow or sweep away immediately. I've been amazed at how quickly an acorn can shoot out a root, necessitating hand removal. This tasty buffet will attract squirrels, resulting in potential digging damage on down the road. The gooey dead blossoms of rhododendrons are yet one more example of maintenance related to vascular plants. Litter patrol is an important duty for moss gardeners.

Dealing with critters

The guest list of creatures drawn to mossy habitats includes a wide variety of species—too many to mention. Get to know the cohabitants of your moss garden by taking time to pause, reflect, and observe. You can saturate your spirit with unforgettable memories. In my moss scape, bunnies, chipmunks, and squirrels scurry to and fro with curiosity and energy. Imagine my delight one afternoon when I watched a squirrel fashion show. A gray squirrel pulled a section of *Thuidium* moss off a wicker chair and wrapped it around his back like a cape!

While you may want a wildlife habitat, be aware that all kinds of birds and small animals may cause damage and necessitate repair of your beloved mosses. Most often the harm is superficial, but repairs can be time consuming as they add up. In spring, birds may fly in to gather moss fragments for use in nests. For me, the biggest issue is when birds are in search of a meal. I have robins and a pair of doves that must have huge appetites and big families to feed. Using netting reduces bird damage and also serves as a deterrent to digging critters such as squirrels, raccoons, and skunks looking for seeds, grubs, and insects. Some moss gardeners report using wire mesh to protect larger moss hummocks.

As an additional solution, scare off nocturnal visitors by installing a motion-sensor light or sonic blaster. An unexpected light or obnoxious noise could very well startle your unwanted guests away. Another preventive method is to sprinkle a hot pepper elixir or bury mothballs to discourage rooting around. By using a chemical grub killer you can eliminate the desired food source for most diggers like raccoons or armadillos. When deer wander in, at least you do not have to worry about them eating up your beauties, but they can mess up mosses with their hoofs. Normally destructive voles and moles do not seem to impact mosses in any negative way.

Although repugnant slugs come out in swarms after dark, mossers do not need to waste time on a slug hunt. Slugs will not devour mosses in the destructive manner in which they annihilate flowers and leaves, at least at my house. (In some other parts of the world, slugs do eat mosses, according to Janice Glime in her online book *Bryophyte Ecology*.) When they scoot across mosses, they deposit a slimy, glimmering trail but do not permanently blemish moss leaves. Because I do not want shiny slug evidence, I just wipe the goop off.

Family pets may be the biggest culprits. Your own dog or cat may cause more damage than all the wild birds and critters put together. One of my friends has playful cats that love pawing and tossing *Leucobryum* colonies in the air. While a sedate dog can walk, sit, and lie around your moss garden without need for concern, a frisky dog is a different matter. I can attest that dogs, oblivious to any obstacles and running at full speed through a moss garden, netted or not, can wreak havoc.

Proof of critter and pet visitation may be apparent in the unwelcome gifts they leave behind. You will definitely want to clean up any feces. Amazingly, mosses do not seem too negatively impacted by dog deposits, but if you leave them for long, they can smother healthy mosses. So whether you find rabbit pellets or dog poop, clean it up. It might be a good idea to rinse off mosses after removing excrement. Regrettably, I must report that male dog urine is harmful to mosses. I have seen definite yellow discoloration and brown spots where male dogs urinate frequently.

The hard work of maintenance tasks should be interspersed with nature breaks. After all, pesky critters are not always up to exasperating mischief, and moss gardens provide a glimpse into the interconnectedness of plants, creatures, and environmental elements. It is fun to bird-watch in your own yard. For that reason, I tolerate a few bird thieves, robins and doves, letting them get away with snatching bits of moss for building nests. Honey bees come to relax and chill out on cool, moist mosses. They do not flit around from plant to plant, blossom to blossom. Instead, bees actually hang out in a stationary position on moss beds for quite a while. These important pollinators will return to the hive to share collected water with their bee buddies. Brilliantly colored butterflies and moths with intricate wing patterns often pause for a drink in my moss garden. I look forward to the migration visits of monarchs each fall when they take a pit stop break from their trek to Mexico.

The wildlife parade continues with beneficial insects living aboveground and underneath mosses. Spiders abound. Thankfully,

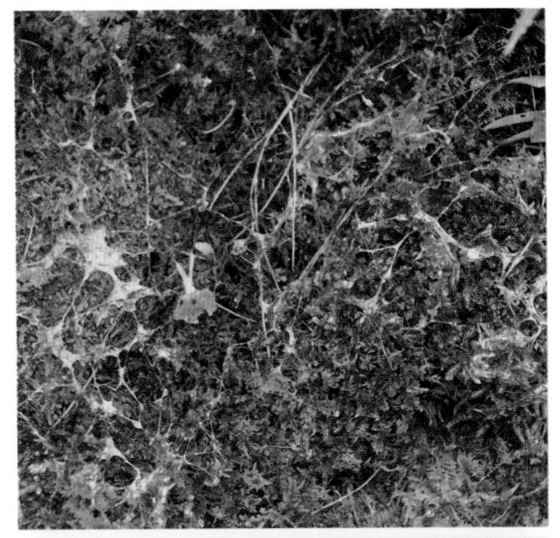

FROM TOP TO BOTTOM
In my garden, slugs cause no permanent damage to mosses even though they do leave a temporary slime trail as evidence of their visit.

Wildlife may leave unwanted gifts like this deer deposit.

Although dog feces do not seem to cause permanent damage, male dog urine can cause major stress and even death to mosses.

my arachnophobia is responding to this moss therapy. I actually enjoy watching hundreds of tiny black spiders with big white butts jumping all around me with their red and black leafhopper friends. Even creepy crawler centipedes have caught my attention, especially the giant yellow-and-black-striped ones. Earthworms thrive in the soil below mosses. These wriggling worms are the tasty morsels that attract many hungry birds. All kinds of invertebrates and microscopic organisms form the population of a moss ecosystem.

The damp environment of a moss garden is inviting to amphibians and reptiles as well. Often frogs surprise me as they plop into a nearby pond from their perch on a mossy log. As might be anticipated, there is a possibility of slithering snakes sneaking around. Be alert to distinguish if you have harmless snakes or dangerous, poisonous species. Know snakes of your region. In America, venomous snakes have triangular-shaped heads with elliptical, not round, pupils. Do not kill snakes when you find them. In the scheme of nature, these reptiles play an integral part in the balance of the overall ecosystem.

Far less threatening and scary, salamanders and newts prefer damp moss homes. Some blend in with surrounding plants while others catch your eye with dazzling colors like orange, red, and yellow. Certain salamanders dart quickly, but some are slow moving. This past summer I had so much fun dancing around with a salamander. It would freeze and become immobile when facing the perceived threat of this moss-dancing fool. It never tried to run away. However, if I got too close, it stopped dead in its tracks—even with a spongy paw held up as if poised to take the next step. It was like a statue until I would move a safe distance away. I provided background music by humming the tune of a Mexican tarantella and waited for it to move again. Although Mr. Salamander was not much of a dance partner, I will treasure the memory of my afternoon frolic with him. As you

maintain your moss garden's beauty, I hope you will take time to let your spirit bond with other living things and fully experience the extraordinary biodiversity of your moss landscape.

Colony repair

Inevitably, you will face a disastrous day when moss colonies are in disarray—upturned, moved out of position, scrunched up, shredded apart. Dislodged colonies could be the result of extreme weather, but usually some wild critter or pet interference is the cause. Whatever the reason, it is a bummer to see your precious mosses tumbled around. With determination, you need to step into the nursemaid role. Pick up any moss debris including colonies and fragments. Make sure to remove any double layers. Turn over any upside-down colonies. Stretch back into place any colonies that look like they were displaced by a paw slide.

As your attention turns to fixing bare spots, loosen the edges of the undisturbed mosses and stretch them out a bit. When moss colonies have decent loft, stretching them back together can be more than enough to fill the holes. Repair any remaining holes from your collection of retrieved colonies or fragments. Fluff up colonies and put them back into place. Add fragments as an extra measure. Water and walk on your repair patches, particularly around the seams or edges. Take advantage of soggy ground after a heavy rainfall by walking repeatedly in troublesome areas to press moss colonies and fragments into the soil.

Netting for protection

I confess that I resisted using netting at first because I considered it obtrusive—taking away from the visual appeal of my moss garden. After having blown and blown, and fixed and fixed many times over, I was willing to give netting a try. An experience with oak catkins (which I refer to in irritation as oak crappers) served as further incentive to explore the advantages of netting. One year in late spring, I had the privilege of leading tours at the Kenilworth moss garden. The day before the event, I expected the incredible green carpet of *Dicranum* and *Hypnum* mosses to greet me. Instead, all I could see was brown covering the ground. To say I was dismayed, or even shocked, would understate my reaction. At first, I thought all the mosses had experienced some intense stress and died. When I realized it was oak litter, I cringed at the thought of such a massive maintenance chore with a pressing deadline. Thankfully, I had reason to jump for joy when the staff gardeners went to corners of the lawn and easily lifted huge sections of netting that had trapped almost all of the dead oak catkins. A quickie blow and the crisis was over in minutes. Whew!

Depending on your perspective—aesthetic, functional, or environmental—netting can be positive or negative. If you do decide to use netting, you can still use a leaf blower to keep your mosses tidy. Be careful not to blow debris under the netting, though. Instead of blowing across the short width, blow lengthwise to avoid trapping leaves underneath.

I use netting only as a temporary solution for dealing with leaves or preventing bird or critter interference. Black wildlife (deer) netting is barely visible from a distance. I prefer using plastic nylon netting with 1-inch holes

Earthworms are common in mosses. These tasty morsels may attract birds, which in turn may dislodge mosses in their quest for food.

Regrettably, you may experience the dismay of dislodged colonies left in disarray by frisky dogs or nighttime critter visitors. Obviously, repairs will be necessary.

that comes on a 7-foot-wide roll in a 100-foot length, which is available at garden centers. Having it on a roll makes placement manageable by one person. However, a second pair of hands is quite helpful. In flat areas, roll it out right on top of mosses, maintaining tautness. In gardens with an uneven landscape, other plants, and/or garden elements, cut the netting to shape around these features.

Wire landscape pins or tent stakes hold the netting down effectively. Unfortunately, metal stakes rust quickly, making it difficult to remove them if you want to lift off the netting. I prefer using hand-sized rocks to anchor corners and hold down edges of the net. Rocks are easy to move around when you want to adjust the netting. If you use care when removing the netting, you can use it again and again. Be forewarned that it tangles easily. If it gets wadded up, recycled netting can be a nightmare. Periodically, you may want to start fresh with a new section from the roll.

If you are dealing with small debris, netting with smaller holes might be a better choice. For instance, it is hard for an acorn to fall through a ½-inch hole. Another netting option promoted by certain moss gardeners is tulle fabric or the netting used for ballerina tutus and square dancing petticoats. With extremely small holes, tulle netting comes in a range of colors. It is available on rolls from fabric stores. I consider tulle fabric far more unattractive than the deer or critter netting options. When wet and glistening from accumulated water beads, tulle jumps out like a sore thumb, but if it is left in place for months or years this visual concern passes after mosses inextricably become embedded in tulle and plant growth rises above the netting.

Botanically speaking, netting is certainly not necessary to establish or maintain colony structure and interconnectedness. Pleurocarpous mosses naturally interconnect, and acrocarpous colonies inextricably bond together anyhow. Netting could be worthwhile if you are attempting soil stabilization or erosion control with mosses. Upright growers can be held together using netting of a smaller gauge than wildlife netting. If you leave any netting in place for very long, healthy mosses will begin to intertwine with each other within the webbing. Once mosses have interconnected with the netting, it is nearly impossible to separate them successfully. Since my use of netting relates to periodic trouble spots and temporary maintenance measures, I do not keep netting in place except for a few weeks at a time. However, some moss gardeners want this extra strength, particularly when remediating erosion issues. Others are convinced netting is integral in growing moss colonies from fragments.

While netting can serve several good purposes, it is not a very natural approach to moss gardening. To me, it seems somewhat incongruous to use such environmentally friendly plants as mosses with a man-made product that will not degrade for decades, maybe centuries. I urge you to consider its value as an interim, preventive measure rather than as a long-lasting

LEFT Netting can be used as a permanent method to hold colonies together or as a temporary measure to deal with critter or bird damage or for seasonal leaf maintenance.

BELOW Stretch nylon netting tightly across your mosses and hold it in place with landscape staples, tent stakes, or rocks in the corners and along the edges.

substrate in your moss gardening efforts. From a tactile point of view, netting reduces the immense pleasure of walking barefoot on your mosses. Instead of enjoying a luxurious experience, you need to walk carefully on a scratchy net so as not to trip and fall.

WINNING THE WAR AGAINST WEEDS

Weeding occupies a high-priority position on the moss maintenance checklist and may be the most time-consuming of all maintenance duties. It is a price you pay for achieving moss excellence. I cannot emphasize enough the need to get rid of weeds in a timely fashion. The longer you wait, the more they spread, and the harder they are to remove. Identifying weeds and getting rid of them from the get-go is the best way to keep invaders at bay. You may wage many battles in your war against weeds. To triumph over aggressive and invasive competitors is a true moss gardening accomplishment.

If weeds already plague your grass lawn and crowd into your flower beds, you can count on dealing with them in your mosses. The "let it grow in" moss gardener may have fewer weeds than the gardener who intentionally introduces mosses. Weeds have a hard time growing in places with nutrient-poor soil, heavy shade, or wet conditions. On the other hand, assertive moss gardeners may have an abundance of weeds. Some of the recommended methods for encouraging moss growth will subsidize healthy weeds. If you plant mosses in the sun, you can expect more weeds than in the shade. If you provide supplemental watering, you may be boosting weed growth as well.

Substrates and weeding

To reduce the need for weeding, you might consider using landscape fabric in the planting phase. Placing a barrier over the soil is particularly advantageous if you are establishing mosses in a former grass lawn location. Mosses can attach to the fabric. Seeds left in the ground from a previous year will not grow up through this barricade. However, many weed seeds fly into nurturing moss colonies via wind dispersal. Therefore, even though weed-inhibiting fabrics hinder growth, weeds still find their way into a moss garden.

Depending on the substrate, your weeding job will be easier or harder. I have found that geotextile fabric works best for me. Tenacious weeds have difficulty permeating this commercial-grade landscape fabric. It is easy-peasy to pull out all sizes of weeds when mosses are growing on this substrate. Additionally, this mat composed of recycled textile fibers maintains its integrity during weed removal—not tearing or stretching out of shape. Conversely, I have gotten really frustrated with other man-made substrates. Weeds effortlessly penetrate felt fabrics—thin hobby kinds and thick, industrial grades. When you try to pull established weeds from felt, the fabric falls apart and stretches out of shape. Other substrates used in my experiments have actually disintegrated.

Netting of any sort makes it more difficult to weed. That is another reason I like to be able to remove netting and not leave it in place forever. Trying to get weeds and their roots through net holes can be a real challenge. Mosses permanently intertwined in netting make weed removal maddeningly difficult. Likewise, it is a lot of trouble to pull weeds out of erosion-control mats made of coir (fibers from coconut husks) and plastic nylon netting. It is imperative to remove intrusive weeds at the earliest stages of growth from netted areas before they get too big.

Weeds to watch out for

Some people say that one person's weed is another's flower, but as far as I am concerned, any weed is a weed in a moss garden. Further, any plant, flower, or tree not intentionally introduced is a target, especially that pervasive one called grass. Which weeds will end up on your removal checklist? Without hesitation, you can probably cite several weeds that are annoyances. In my case, dandelions and Japanese stiltgrass persistently push into moss areas each year. Since weeds vary geographically, gaining a better understanding of common weeds and invasive plants in your region is worthwhile.

From a global perspective, plain ol' weeds are most annoying since many species grow quickly, sometimes reaching maturity and reseeding within a few days. Individual weeds can produce mass quantities of seeds; for instance, crabgrass can have up to 150,000 seeds on each plant. Horizontal spreaders are especially aggravating, winding underneath mosses with alarming spread. New growth can pop up from nodes along creeping stems of weeds such as white clover.

Beyond typical weeds, you may face non-native invasive plants. Many plants originating in Asia and Europe and popularly used in American gardens for decades are now problematic. At the very minimum, keep any invasives sequestered from moss garden areas. Officially, native plants are not invasive. Yet, some indigenous plants definitely have aggressive tendencies, particularly in a moss garden. Jewelweed grows like a weed in my yard. Corkscrew seedpods explode in all directions, inspiring its other common name, touch-me-not. Each spring, thousands of baby jewelweeds shoot up in my moss lawn. It is absolutely imperative for me to address this mass attack before they grow to full maturity. Although roots are shallow, larger plants have a horizontal, round ball-like root mass that makes a significant hole (4 to 5 inches across) when pulled out. I let them grow in another area near an alcove of azaleas. By confining jewelweed, I still get to enjoy darting hummingbirds and to have a convenient salve when I get stung.

In your moss maintenance process, you likely will become acquainted with small-scale weeds that might have gone unnoticed in a grass lawn. Recognizing all weed species is a valuable skill. Learn to spot tricksters like birdseye pearlwort (*Sagina procumbens*). This weed looks a lot like a moss. Its leaf arrangement reminds me of *Polytrichum*. Single seedpods resemble spore capsules, but white flowers are a dead giveaway that this imposter is not a true bryophyte. The shallow horizontal roots of this perennial weed exhibit the willpower of a stubborn mule. Immediate removal is required to prevent quick-spreading growth. It tops my list of obnoxious weeds in my moss garden.

Do not forget to weed out renegade tree seedlings such as oaks, pines, and maples. When young and cute, they complement the miniature aspects of a moss garden. But these babies will grow up quickly, and roots can be strong. Resist leaving them for very long.

Learning to identify bothersome weeds specific to your locale is vital to your moss gardening success. The examples I have provided

While tulle mesh or deer netting may reduce critter damage, pulling weeds out may be more difficult and a pain to accomplish. The smaller holes of tulle netting make it really hard to get the weeds out without destroying the mesh fabric.

are just the beginning. I suggest you visit Internet sites, your local library, or the cooperative extension office as you develop the skill of recognizing weeds.

On a final note about what to weed, you may need to weed stray mosses out of mosses. Species that drift into other species can add interest. It is cool to see unexpected little moss trees emerging from pleurocarpous carpets or acrocarpous mounds—the results of successful spore dispersal and vegetative reproduction. However, aggressive tendencies of certain mosses may necessitate their removal if you want to maintain the integrity of your design or a consistent texture.

Weeding methods

With the glut of aggressive weeds and undesirable plants that can potentially move into your moss garden, elimination is a primary concern. Hand weeding is the most eco-friendly way to get rid of them. Tiny roots may extend further into the ground than you realize and the weeds might reappear. Take time to weed with intentional care. If weedy plants have gone to seed, cup your hand around the pod to prevent seeds from dispersing during removal tasks. Eventually, many aspects of your moss-weeding skills will improve, including speed and comfort level. I am happy to say that I am now ambidextrous, with the ability to weed with either hand. I'm a "switch-weeder"!

To minimize damage to mosses during the removal process, special techniques are required that are different from typical weeding practices. You cannot yank weeds out or you might mar the attractiveness of mosses during the process. To me, hand weeding is just that—using your hands and developing a special touch to achieve your goal of weed eradication. My favorite moss-weeding method calls for hands first, with tools as a last resort. When absolutely necessary, sharp garden tools, a small knife, or even tweezers might come in handy for really stubborn weeds. It is better to weed when the ground is wet and far more difficult to remove weeds successfully when the soil is hard packed and dry.

Make yourself comfortable. As with planting mosses, you may choose to sit or kneel—whichever suits you. Wear loose clothing that will allow a full range of movement so that you can stretch out easily. Weeding can serve as your exercise session for the day. Once I am comfortably in place, I weed within my entire reachable circumference, stretching as far as I can. From a seated position, I pirouette around using all kinds of unused muscles. While sitting is my normal position, I sometimes lie down to weed on my side or my stomach to maximize my coverage before standing up and moving to the next spot. When your face is that close to the ground, you can search out the base of tiny weeds to get a good hold.

It is best to use two hands in the weeding process. One hand holds mosses in place while you pull weeds with the other hand. With one hand, spread your fingers over the mosses and fork them around the weed. With your primary hand, find the base of the weed. Gently use a circular motion to wriggle it out.

Your methods for removing weeds from moss colonies depend on whether the weed species has a horizontal root system, a taproot, or both. In general, upright weeds are easier to remove from mosses than horizontal growers. You can take hold of a single upright weed and pull it out vertically using back and forth swaying and tugging motions. If the upright weed has a strong taproot (for instance, dandelion), you will need to pull hard to get it all the way out. If you break a tough root off, you can use a tool or knife to dig out the remainder. If you dig deeply, keep your hole narrow. You do not want to leave wide, gaping holes.

Plants with horizontal root systems and stolons can branch out in many directions. Attached to soil near the surface, these weeds require special attention and meticulous care. Find the mother plant and use both hands to

TOP Be observant of tiny weeds in your maintenance patrols. The aggressive weed in the center, birdseye pearlwort (*Sagina procumbens*), looks a lot like a moss, doesn't it?

Often you will find that different moss species will cohabitate. Be aware that some species may be more aggressive growers, and you may decide to weed these out to maintain the integrity of the colony. Or simply enjoy the combination.

Make sure to pull out horizontal roots with a delicate touch. I have found that wearing garden gloves impedes my ability to feel when they release. Upon occasion, a knife or weed tool may be necessary for taproots.

loosen it. Further, follow the trails above and/or below mosses to release root systems. This is a time when gloves definitely hamper your ability to distinguish and grab trailing roots. You will need to fish around underneath pleurocarpous mosses to feel directional growth and to capture roots of each attached node. Once you get the roots loose, while keeping your second hand on top of the mosses, gradually and tenderly pull out the roots from one direction while still underneath the mosses. This sideways method is more precise and less damaging than pulling through the top. If you create a hole when weeding, gently tuck mosses back in place or stretch adjacent mosses together. You may not even be able to see where the weed once lived.

Despite your best efforts to keep moss damage to a minimum, some fragments will stay attached to the roots of confiscated weeds. Again, your second hand comes in handy. After the weed is out, get in the habit of immediately using a quick retrieval stroke to slide moss fragments out of root systems. If you are stressing over throwing away any moss morsels, hoard them in a separate container. Of course, you benefit from stockpiling moss fragments for "overseeding" in spots that are bare.

Eradicate poison ivy as soon as you spot the leaves of three, which are simple to recognize. In winter months, be careful when pulling roots that are reddish or hairy. The urushiols (poisons) can cause rashes when the plants and vines are dormant, even dead. Gloves are advised whether you think you are allergic or not. You never know when your sensitivity will change. Often, poison ivy is intertwined within rescued moss colonies, so beware.

Hand weeding affords an incredibly intimate opportunity to connect with your mosses. Try to take the chore out of weeding by enjoying the minute aspects of bryophytes. One day, after hours of methodical weeding, I came across a surprise. While pinching out that exasperating birdseye pearlwort, I noticed small sections of dark green leaves and hornlike protrusions. To my immense delight, it turned out to be a hornwort, *Anthoceros carolinianus*. Knowing there are approximately one hundred species of hornworts in the world, compared to thousands of mosses and liverworts, I was ecstatic to find hornworts growing in my moss garden. Keep your close-up lens within easy reach so you can get an eye-opening perspective. Take weeding breaks to relax and embrace your world.

Using herbicides

As a moss gardener, you do *not* want to pay the price for *not* keeping up with weeds in your maintenance routine. Weeds can get out of control before you know it. Some landscapers and moss gardeners report successful results in using chemical or organic herbicides to control weeds in mosses as an alternative to the labor-intensive method of weeding by hand. The application of herbicides can result in widespread death of weeds, yet, thankfully, not mosses. Without a fully developed vascular system, mosses should not die from a systemic poison.

Products widely used by home gardeners are numerous and marketed under a barrage of labels. Chemical herbicides are either selective (plant-specific) or nonselective (killing all vegetation). If weeds are already present, postemergence herbicide is required. Many postemergence chemicals are effective with broadleaf plants (a botanical term that refers to dicot properties, not big leaves). If you want to prevent weeds from starting, you use a preemergence type. Day or night temperatures along with germination times and growth stages are factors to be considered with preemergence herbicides—usually applied in spring and fall.

You need to know your weeds before choosing a chemical herbicide. Certain pernicious weeds, like crabgrass, require spot treatment. Contact poisons, which affect only parts of the plant they touch, are good for annuals. Systemic killers effectively translocate through the plant to underground roots and stolons of perennial weeds or plants. Glyphosate-based herbicides can be used in a 1:10 mixture (one part chemical, ten parts water) without killing mosses. Weed- and grass-killer products that include the active ingredients diquat dibromide, fluazifop-p-butyl, and dicamba require no dilution.

To build your confidence level with this chemical treatment method, try a small trial section first before applying chemicals to extensive areas. After weeds wilt, add one more vascular plant maintenance task—removing dead foliage. Occasionally, certain moss species may show temporary signs of stress by displaying dingy, yellowish foliage or black stems. A vigorous fluff should alleviate this stressed appearance. Although weeds will die in a matter of hours or days, it might take mosses weeks to display any warning signs. Obviously, discontinue use if any moss trauma persists.

On an environmental note, chemicals may cause harm to other plants or animal life as well as impact groundwater quality. Opponents of excessive chemical use in agriculture also warn of toxic dangers to your personal health. Consider wearing a filter mask to protect your lungs, and avoid spraying on a windy day. Let your conscience be your guide regarding environmental impact or health hazards related to using garden chemicals. It is not comforting to me that previously customary practices are now considered dangerous and that once-popular chemicals have been taken off the market. It is scary to me that herbicides are available today that contain 2,4-dichlorophenoxyacetic acid (known as 2,4-D)—the active ingredient in Agent Orange.

If you want to avoid chemicals, try organic compounds or mixtures. Unfortunately, after some experimentation, I have discovered that agricultural vinegar (20 percent) is *not* a good alternative. *Atrichum* mosses receiving the chemical weed and grass killer product did just fine. Weeds died. Mosses were unstressed. However, the strong vinegar annihilated all mosses along with the weeds—at least for a few months. It took six months for the mosses to recover with any new growth.

Always read instructions and follow precautionary human safety measures when applying

Rubbing your hands over mosses—fluffing—helps perk them up. If you experience any mold or fungus, fluff vigorously to remove the threat; let colonies dry out a bit. Most likely, colonies will recover.

herbicides to kill weeds and grasses growing in your mosses. Avoid windy days and try to treat when no rain is predicted for twenty-four hours. Also, turn off the watering system and postpone any watering activities until after herbicide application. If you are using granular preemergence products, water in chemicals within one to three days to help granules dissolve. Expect best results when weeds are young. Multiple applications may be necessary throughout the entire season as part of your weed-control measures.

Fluffing and trimming for rejuvenation

While weeding, you can also fluff and hand groom your mosses. As a touchy-feely kind of person, I like to touch my mosses. Sweeping a hand across soft mosses is such a pleasure. You can justify this indulgence by delicately raking your fingers through your mosses to dislodge loose gametophytes, remove dead portions, and freshen up the colony. Gather fragments from the top of the colony like bread crumbs off a table. Using your small whisk broom, you can sweep leaf tips from *Leucobryum* colonies.

You might even grab your scissors and carefully snip out ugly gametophytes for a more pleasant look. You can also give mature moss colonies a haircut. Trimming the tops of plants freshens the colony and stimulates new growth (like pruning shrubs). Some moss species (such as *Polytrichum* and *Climacium*) have plant stems that are wiry and strong. Cut off sections ½ to 1 inch long.

All of these techniques yield fragments to add to your propagation pile. Even dead-looking moss fragments might generate new growth—so do not discard them. Instead, plant the pieces in new places or use them as "seed" plants.

TOP Give your mosses a haircut with scissors to freshen the colony. Within a couple of months, colonies will be covered with new growth.

Brown sections of *Polytrichum commune* are the older growth that can be snipped off. Note new green growth appearing from tips.

THE ROAD TO MOSS GARDENING SUCCESS

We have come to the end of this book, but I hope it will be the beginning (or continuation) of your own moss journey. We have covered a wealth of information about moss gardening that provides you with a basic understanding of how to proceed. As you move forward with mosses in your landscape, I wish you success. May your spirit soar in your moss retreat and your connection to nature be enhanced. May you experience the exhilaration of a triumphant moss gardener as you gaze upon your expanses of green. May your feet delight in the massaging comfort of a moss lawn. Throughout all the seasons, may you enjoy the benefits of your eco-friendly mosses and take pride in your role as a responsible and caring steward of our environment.

I encourage you to share your experiences—both successes and failures—with fellow moss enthusiasts so that we can build a valuable body of knowledge together. Social networking sites and garden blogs offer a forum for exchange. Visit my Web site (mountainmoss.com) and my Facebook group, Go Green With Moss, to connect with mossers from around the world.

In conclusion, I hope you will revel in your ability to create resplendent moss gardens that will delight your family and friends for years to come. And keep on mossin'.

The end . . . of this book. But I hope it is the beginning (or continuation) of your own moss journey. Rest assured, I will keep on truckin' . . . and mossin'!

OPPOSITE I expect my moss garden to continue to provide years of pleasure. Don't you want a moss garden, too?

metric conversions

inches	cm	mm
1/32		1
1/16	0.2	1.6
1/8	0.3	3.2
1/4	0.6	6.4
3/8	0.9	9.5
1/2	1.3	12.7
5/8	1.6	
3/4	1.9	
7/8	2.2	
1	2.5	
2	5.1	
3	7.6	
4	10	

feet	meters
1	0.3
2	0.6
3	0.9
4	1.2
5	1.5
10	3

1 acre .... 0.405 hectare
1 gallon (U.S.) .... 0.833 British gallon
1 ton (2000 lbs.)..... U.S.0.907 metric ton

Temperatures

°C = 5/9 × (°F – 32)
°F = (9/5 × °C) + 32

resources and references

If you would like to learn more about mosses and their value to our environment and in your garden, please visit any of the Web sites and/or read any of the books listed here. If you want to gain a more scientific perspective, Janice Glime's ebook, *Bryophyte Ecology*, tops the list of excellent references. To purchase live mosses for your landscape, I hope you will consider the quality products offered by Mountain Moss Enterprises, my business, at **mountainmoss.com**. To engage my services as a moss garden designer, consultant, landscaper, or lecturer, please contact me directly: mossinannie@gmail.com.

Favorite Web sites

Annette Launer's German-language site, Bryophyta, bryophyta.pflanzenliebe.de/moose_impressum.html

Ann Franck Gordon's site, Moss Whispers, mosswhispers.com/

Australian National Herbarium, Australian Bryophytes, anbg.gov.au/bryophyte/index.html

Brian Engh's video, Moss: A Tribute, youtube.com/watch?v=MWRazeUhg44

Bryophyte Flora of North America, mobot.org/plantscience/bfna/bfnamenu.htm

Duke University Herbarium, biology.duke.edu/herbarium/databasing.html

Eagle Hill Institute, eaglehill.us/

Janice Glime's Bryophyte Ecology site, bryecol.mtu.edu

Highlands Biological Station, highlandsbiological.org/

Oregon State University, Living with Mosses page, bryophytes.science.oregonstate.edu/mosses.htm

Southern Illinois University, Carbondale, Bryophytes page, bryophytes.plant.siu.edu/index.html

United States Department of Agriculture, Natural Resources Conservation Service, PLANTS Database, plants.usda.gov/

University of California, Berkeley, California Moss eFlora, ucjeps.berkeley.edu/CA_moss_eflora/

University of Wisconsin, Stevens Point, Robert W. Freckmann Herbarium, wisplants.uwsp.edu/

Recommended books

Breen, Ruth S. 1963. *Mosses of Florida*. Gainesville, FL: University of Florida Press.

Crum, Howard, and Lewis Anderson. 1981. *Mosses of Eastern North America*. 2 vols. New York: Columbia University Press.

Davison, Paul. 2008. *A Trailside Guide to Mosses and Liverworts of the Cherokee National Forest*. San Francisco: Blurb Books.

Dillenius, Johann Jakob. 1768. *Historia Muscorum: A General History of Land and Water*. Reprint, Charleston, SC: Nabu Press, 2011.

Flora of North America Editorial Committee, eds. 2007. *Bryophyta: Mosses, Part 1*. Vol. 27 of *Flora of North America North of Mexico*. New York and Oxford: Flora of North America Association.

Glime, Janice M. 1963. *The Elfin World of Mosses and Liverworts of Michigan's Upper Peninsula and Isle Royal*. Houghton, MI: Isle Royale Natural History Association.

———. 2006–2015 (ongoing). *Bryophyte Ecology*. 5 vols. Ebook sponsored by Michigan Technological University and the International Association of Bryologists. bryoecol.mtu.edu/.

Goffinet, Bernard, and A. Jonathan Shaw. 2008. *Bryophyte Biology*. 2nd ed. New York: Cambridge University Press, 2000.

Grout, A. J. 1940. *Moss Flora of North America*. 3 vols. Reprint, New York: Hafner Publishing, 1972.

Kimmerer, Robin Wall. 2003. *Gathering Moss: A Natural and Cultural History of Mosses*. Corvallis, OR: Oregon State University Press.

Malcolm, Bill, and Nancy Malcolm. 2000. *Mosses and Other Bryophytes: An Illustrated Glossary*. 2nd ed. Nelson, New Zealand: Micro-Optics Press.

Malcolm, Bill, Nancy Malcolm, Jim Shevock, and Dan Norris. 2009. *California Mosses*. Nelson, New Zealand: Micro-Optics Press.

McKnight, Karl B., Joseph R. Rohrer, Kirsten McKnight Ward, and Warren J. Perdrizet. 2013. *Common Mosses of the Northeast and Appalachians*. Princeton, NJ: Princeton University Press.

Munch, Susan. 2006. *Outstanding Mosses and Liverworts of Pennsylvania and Nearby States*. Mechanicsburg, PA: Sunbury Press.

Richards, Paul. 1950. *A Book of Mosses*. Middlesex, England: King Penguin.

Schenk, George. 1997. *Moss Gardening, Including Lichens, Liverworts, and Other Miniatures*. Portland, OR: Timber Press.

Vitt, Dale H., Janet E. Marsh, and Robin B. Bovey. 1988. *Mosses, Lichens, and Ferns of Northwest North America*. Reprint ed. Vancouver, British Columbia: Lone Pine Publishing.

Weber, William, and Ronald Wittmann. 2007. *Bryophytes of Colorado: Mosses, Liverworts, and Hornworts*. Santa Fe, NM: Pilgrims Process.

acknowledgments

My moss journey has filled my spirit with rejuvenating joy. Days spent exploring the woods, creating moss gardens, and even weeding are treasured memories shared with family, friends, and co-workers. I am grateful to a multitude of folks who have cheered me on and encouraged me to forge ahead with my moss ventures and to follow my dreams despite obstacles along the way. I shall ever be indebted to new and old moss friends who have been by my side through all the highs and lows of growing a moss business. Writing this book has been a monumental task and the culmination of hard work. It is not a solitary process but instead has represented the true sense of a village. Thanks, y'all! This book would not have materialized without your guidance and support.

Always in my corner and my closest moss companions, my brother, Dale Martin, and my best friend, Leila von Stein, know better than anyone else the challenges and rewards of mossin'. Countless hours spent together have been my privilege. Thank you for always helping me and running to my rescue, no matter what. My sons, Flint Martin Barrow and Carson Martin Escoe, have contributed immensely. Carson has helped me at the mossery, assisted with moss rescues, and offered good garden design advice. Both boys assisted in the transformation of the asphalt driveway into a marvelous moss garden. The success of my Web site is due to Flint's savvy computer skills. My other "brother," James Neely, continues to provide moral support and never fails to lend a helping hand. Thank you just doesn't seem like enough to say. The influence of my mother, Rachel Baker Martin, and my sister, Kay Martin Jones, has spurred me to believe in myself since childhood. And I can thank Daddy for instilling the desire to always pay attention to detail, encouraging me to maintain quality standards, and giving me the gumption to follow my dreams.

I extend particular thanks to Bill Layton, who has been my steadfast business advisor for years now. His professional advice, keen editing skills, and continued friendship are key factors in arriving at this pinnacle in my moss career. In a similar manner, Dr. Janice Glime has graciously shared her incredible expertise as a bryologist. While we have never met in person, this new moss friend has been a jewel while I have written this book. These two, along with Susan Sunflower (Master Gardener friend) and Bea Gardner Mattice (childhood friend), have donated many hours reviewing my book and offering suggestions on how to improve it.

I owe so much to friends who have shared their property with me to start and continue to grow my mossery operations. Betsy Smith in Penrose along with Claire and Eric Stephenson in Cedar Mountain have been essential in the process of establishing my cultivation activities. Their generosity of spirit will always be remembered. In addition, over the past few years a number of people have assisted in my mossery operations. Special thanks to Sara Boggs, Jason Hughes, Dean Hughes, James Wood, Perry King, Susan Johnson, Tina Watkins, Amy Jo Riddick, and Joe Bruneau. As I move forward at my new mossery, I appreciate the support of Bruce and Belinda Roberts as well. The camaraderie of hard work creates bonds that will always last. A hearty call-out to all Mountain Moss customers, especially Ben Carter, for offering opportunities to showcase my mossin' talents. I also appreciate the encouragement and support of friends met through educational workshops and social networking sites on the Internet, particularly my bryologist friends on Bryonet.

It has been a pleasure to get to know other mossers from near and far. I extend my thanks to each one of the profiled moss gardeners for their time and willingness to impart their experiences in this book. I appreciate Robert Balentine, Norie Burnet, John Cram, Ann Gordon, Dale Sievert, and George and Carol Vickery for sharing their moss joys with me and adding their perspectives about moss gardening.

The spirit of worldwide community has been reinforced by the contributions of numerous photographers who shared their superb images to be featured in this book. I have been impressed with the willingness of these talented photographers to share their talents. Once again, I am grateful to people I've never even met who have been willing to help me out. From Japan to New York to the West Coast, photographers have agreed to provide images. Special thanks to Bill and Mardy Murphy for shooting my portrait and the portraits of the main characters in this book, the mosses! They are the ringleaders, but numerous other photographers deserve my gratitude: Graham Bell, Des Callahan, Joseph Cooper, Michael DeMeo, Hinonori Geduchi, Tony Giammarino, Janice Glime, Robin Haglund, Taylor Ladd, Annette Launer, Xi Lhing, J. Paul Moore, Geert Raeymaekers, Mark Schueler, Blanka Shaw, Dale Sievert, and James Wood.

I am indebted to my editors at Timber Press. Without their professional guidance, this book would not have been possible. I so appreciate the opportunity they have given me to share my passion for moss gardening with others. I extend my sincerest thanks to Juree Sondker, Sarah Milhollin, Eve Goodman, and Andrew Beckman. Most of all, I am grateful to Lorraine Anderson, an "eco-sister" with an understanding of the connection of nature to our spirits, who has provided exceptional editorial assistance and professional guidance in preparing this book for publication.

Each and every one of you has been among my villagers. This mosser is one happy camper thanks to all your support. Many thanks for contributing to my moss journey!

photo credits

Courtesy of Graham Bell, Butchart Gardens, page 46

Joseph Cooper, page 50

Michael DeMeo, moss and terrarium enthusiast, page 161

Hinonori Geduchi, Digital Natural History Museum of Hiroshima University, pages 33, 34, 36 left, 38, 39, 40

Tony Giammarino, pages 53, 137, 144 bottom, 173–175

Janice M. Glime, pages 6, 35, 36, 85

Paul Jones, photo courtesy of Sarah P. Duke Gardens, page 47

Taylor S. L. Ladd, pages 13, 92, 93, 122, 193

Annette Launer, pages 6, 25, 84, 94, 129 left

Sarah Milhollin, pages 42, 43

J. Paul Moore, page 165

Mardy Luther Murphy, page 188 left, 240

William J. Murphy, pages 1, 6–7, 8, 30, 70, 73 bottom, 75, 79, 80 bottom, 82 bottom, 86, 87, 100, 103–116, 118–128, 129 right, 130 top, 131–134, 162, 190

Mary Beth O'Connor, Associate Professor of Communication, Purdue University Calumet, page 158 top

Geert Raeymaekers, Brussels (BE), page 49

Mark Schueler, page 159

Blanka Shaw, page 28

Dale Sievert, pages 4–5, 36 right, 73 top, 80 top, 81, 146–149

James Wood, page 45

© Li Zhang, page 170 bottom

All other photos are by Annie Martin

index

A

B

C

G

H

M

N

O

P

Q

R

S

T

U

V

W

Y

Z

about the author

©Mardy Luther Murphy

Fascinated with mosses since childhood, Annie Martin has earned her reputation as an expert through years of personal research and experimentation with moss gardening methods. Owner of Mountain Moss Enterprises, Mossin' Annie wears many moss hats: plant rescuer, farmer, landscaper, lecturer, educator, and now, author. Born and raised in North Carolina's Blue Ridge Mountains, she passionately advocates the advantages of eco-friendly mosses by imparting how-to knowledge from a practical point of view. Furthermore, she emphasizes environmental benefits to our world while placing value on the interconnectedness of nature and the human spirit. To continue your journey through the magical world of mosses, visit her Web site: **mountainmoss.com.**